124
Advances in Biochemical Engineering/Biotechnology

Series Editor: T. Scheper

Advances in Biochemical Engineering/Biotechnology

Series Editor: T. Scheper

Recently Published and Forthcoming Volumes

High Resolution Microbial Single Cell Analytics
Volume Editors: Müller S., Bley, T.
Vol. 124, 2011

Bioreactor Systems for Tissue Engineering II
Strategies for the Expansion and Directed Differentiation of Stem Cells
Volume Editors: Kasper C., van Griensven M., Pörtner, R.
Vol. 123, 2010

Biotechnology in China II
Chemicals, Energy and Environment
Volume Editors: Tsao, G.T., Ouyang, P., Chen, J.
Vol. 122, 2010

Biosystems Engineering II
Linking Cellular Networks and Bioprocesses
Volume Editors: Wittmann, C., Krull, R.
Vol. 121, 2010

Biosystems Engineering I
Creating Superior Biocatalysts
Volume Editors: Wittmann, C., Krull, R.
Vol. 120, 2010

Nano/Micro Biotechnology
Volume Editors: Endo, I., Nagamune, T.
Vol. 119, 2010

Whole Cell Sensing Systems II
Volume Editors: Belkin, S., Gu, M.B.
Vol. 118, 2010

Whole Cell Sensing Systems I
Volume Editors: Belkin, S., Gu, M.B.
Vol. 117, 2010

Optical Sensor Systems in Biotechnology
Volume Editor: Rao, G.
Vol. 116, 2009

Disposable Bioreactors
Volume Editor: Eibl, R., Eibl, D.
Vol. 115, 2009

Engineering of Stem Cells
Volume Editor: Martin, U.
Vol. 114, 2009

Biotechnology in China I
From Bioreaction to Bioseparation and Bioremediation
Volume Editors: Zhong, J.J., Bai, F.-W., Zhang, W.
Vol. 113, 2009

Bioreactor Systems for Tissue Engineering
Volume Editors: Kasper, C., van Griensven, M., Poertner, R.
Vol. 112, 2008

Food Biotechnology
Volume Editors: Stahl, U., Donalies, U. E. B., Nevoigt, E.
Vol. 111, 2008

Protein-Protein Interaction
Volume Editors: Seitz, H., Werther, M.
Vol. 110, 2008

Biosensing for the 21st Century
Volume Editors: Renneberg, R., Lisdat, F.
Vol. 109, 2007

Biofuels
Volume Editor: Olsson, L.
Vol. 108, 2007

Green Gene Technology
Research in an Area of Social Conflict
Volume Editors: Fiechter, A., Sautter, C.
Vol. 107, 2007

Cell Separation Fundamentals, Analytical and Preparative Methods
Volume Editors: Kumar, A., Galaev, I.Y., Mattiasson, B.
Vol. 106, 2007

High Resolution Microbial Single Cell Analytics

Volume Editors:
Susann Müller · Thomas Bley

With contributions by

M. Berney · L. M. Blank · T. Bley · P. Bombach · C. L. Davey ·
H. M. Davey · T. Egli · I. Fetzer · R. Geyer ·
J. A. J. Haagensen · F. Hammes · H. Harms ·
T. Hübschmann · S. Jakobs · S. Kleinsteuber ·
H. Kortmann · S. Müller · D. Neumann · B. Regenberg ·
A. Schmid · J. D. Stenson · C. Sternberg · S. Stoldt ·
C. R. Thomas · K. Yasuda · Z. Zhang

Springer

Editors
PH Dr. Susann Müller
Helmholtz-Zentrum für
Umweltforschung
GmbH (UFZ)
Dept. Umweltmikrobiologie
Permoserstr. 15
04318 Leipzig
Germany
Email: susann.mueller@ufz.de

Prof. Dr. Thomas Bley
TU Dresden
Fak. Maschinenwesen
Inst. Lebensmittel- und
Bioverfahrenstechnik
Bergstr. 120
01069 Dresden Sachsen
Germany
Email: thomas.bley@tu-dresden.de

ISSN 0724-6145
e-ISSN 1616-8542

ISBN 978-3-642-26732-1
ISBN 978-3-642-16887-1 (eBook)

DOI 10.1007/978-3-642-16887-1

Springer Heidelberg Dordrecht London New York

Softcover reprint of the hardcover 1st edition 2011

Cover design: WMXDesign GmbH, Heidelberg, Germany

Printed on acid-free paper.

Springer is part of Springer Science+Business Media (www.springer.com)

Series Editor

Prof. Dr. T. Scheper

Institute of Technical Chemistry
University of Hannover
Callinstraße 3
30167 Hannover, Germany
scheper@iftc.uni-hannover.de

Volume Editors

PH Dr. Susann Müller

Helmholtz-Zentrum für
Umweltforschung
GmbH (UFZ)
Dept. Umweltmikrobiologie
Permoserstr. 15
04318 Leipzig
Germany
susann.mueller@ufz.de

Prof. Dr. Thomas Bley

TU Dresden
Fak. Maschinenwesen
Inst. Lebensmittel- und
Bioverfahrenstechnik
Bergstr. 120
01069 Dresden Sachsen
Germany
thomas.bley@tu-dresden.de

Editorial Board

Honorary Editors

Advances in Biochemical Engineering/Biotechnology

Advances in Biochemical Engineering/Biotechnology is included in Springer's ebook package *Chemistry and Materials Science*. If a library does not opt for the whole package the book series may be bought on a subscription basis. Also, all back volumes are available electronically.

For all customers who have a standing order to the print version of *Advances in Biochemical Engineering/Biotechnology*, we offer free access to the electronic volumes of the Series published in the current year via SpringerLink.

If you do not have access, you can still view the table of contents of each volume and the abstract of each article by going to the SpringerLink homepage, clicking on "Chemistry and Materials Science," under Subject Collection, then "Book Series," under Content Type and finally by selecting *Advances in Biochemical Bioengineering/Biotechnology*

You will find information about the

- Editorial Board
- Aims and Scope
- Instructions for Authors
- Sample Contribution

at springer.com using the search function by typing in *Advances in Biochemical Engineering/Biotechnology*.

Color figures are published in full color in the electronic version on SpringerLink.

Aims and Scope

Advances in *Biochemical Engineering/Biotechnology* reviews actual trends in modern biotechnology.

Its aim is to cover all aspects of this interdisciplinary technology where knowledge, methods and expertise are required for chemistry, biochemistry, microbiology, genetics, chemical engineering and computer science.

Special volumes are dedicated to selected topics which focus on new biotechnological products and new processes for their synthesis and purification. They give the state-of-the-art of a topic in a comprehensive way thus being a valuable source for the next 3–5 years. It also discusses new discoveries and applications.

In general, special volumes are edited by well-known guest editors. The series editor and publisher will however always be pleased to receive suggestions and supplementary information. Manuscripts are accepted in English.

In references *Advances in Biochemical Engineering/Biotechnology* is abbreviated as *Adv. Biochem. Engin./Biotechnol.* and is cited as a journal.

Special volumes are edited by well-known guest editors who invite reputed authors for the review articles in their volumes.

Impact Factor in 2009: 4.165; Section "Biotechnology and Applied Microbiology": Rank 23 of 150

Attention all Users of the "Springer Handbook of Enzymes"

Information on this handbook can be found on the internet at springeronline.com

A complete list of all enzyme entries either as an alphabetical Name Index or as the EC-Number Index is available at the above-mentioned URL. You can download and print them free of charge.

A complete list of all synonyms (more than 57,000 entries) used for the enzymes is available in print form (ISBN 978-3-642-14015-0) and electronic form (ISBN 978-3-642-14016-7).

Save 15%

We recommend a standing order for the series to ensure you automatically receive all volumes and all supplements and save 15% on the list price.

Preface

Cells are individuals. Even microbial cells, which in case of some environmental bacteria present a size down to 0.4 or 0.5 μm, deviate from each other in terms of life cycle, protein composition, and metabolism. This heterogeneity results from distinct intrinsic cell features like age, stage in cell cycle and position of the division plane, gene transfer or loss, mutations, or epigenetic inheritance. Similarly, external parameters influence cellular features due to various micro-environmental inhomogeneities comprising, e.g., the availability of carbon or other sources of energy and of electron acceptors like oxygen and the prevalence of stress conditions like, e.g., mechanical pressure.

All these parameters influence the efficiency of a cell in a biotechnological process. Since every cell contributes to the product yield of, e.g., a fermentation process in industrial biotechnology, dead, inactive or weakly active cells will limit this productivity. Therefore, single cell-related analytical techniques need to be involved in the evaluation and control of such processes. Although many of such technologies are already successfully used in medical sciences, where human cell populations are investigated with a new generation of amazing instruments, this is not true for microorganisms. One reason for this discrepancy is the priority of medical research in terms of funding. A more scientific point is the fact that microbial cells comprise only a thousandth of the volume of a normal blood cell and are therefore much more difficult to observe and to analyse.

In recent years, however, microorganisms have started to come into the focus of many fields as, e.g., chemistry, which were since long thought to have no interest in these 'un-steerable' organisms. This is because microorganisms are not only tremendously diverse from a phylogenetic point of view, they also catalyse a wealth of biochemical processes which can be used, e.g., in white biotechnology or in energy producing processes like biogas production. Very often the organisms involved in such applications are still unknown with regard to affiliation and function. Since most of these microorganisms still cannot be cultivated as a pure culture, single cell techniques to follow their performance are of utmost interest and necessity.

Still such techniques are expensive and often difficult to operate. Usually they are used in research laboratories to understand the very basic principles of microbial life. In a few cases, however, people have already tried to obtain information referring to individual microorganisms by using single cell technologies, either relying on sophisticated but also on cheaper equipment, based on chip- and microfluidic devices. Remarkable insights into cell behaviour have already been obtained by using such small-scale and sometimes partly disposable instrumentations.

Population dynamics or subpopulation dynamics in biotechnologically or environmentally relevant processes are responsible for the variability in seemingly homogeneous populations under seemingly homogeneous micro-environmental conditions and result in surprisingly quick intra-population changes within a 'stable' process. Here, not only live/dead states play a crucial role in population development but also the above mentioned intrinsic parameters. For a deeper understanding and forecasting of the behaviour of microbial populations quantitative analysis and mapping into mathematical models will provide the indispensable theoretical foundations. In this context, we can make use of a broad panel of different model concepts. Their usefulness has to be proven by their ability to assign the characteristic features of single cells or of segregated subpopulations to the model variables. These models will ultimately allow to develop, control and enhance microbial performances in bioreactors or in locally confined, natural systems where microorganisms are used for distinct tasks.

All in all, microbial single cell analytics evolved to a large degree within the last 10 years. Nevertheless, these technologies are still on the edge and have the potential to become far more usable and useful for basic research and for application in already well established microbial processes. We hope that this volume intrigues the reader to learn more about microorganism and their complexity but mainly on the techniques which can be used to understand their basic principles of live and survival. Highly resolved information on these small organisms will enable us to quantify their life and to orchestrate their abilities to a successful control and optimisation of bioprocesses.

We would like to thank all the authors for their valuable contributions and discussions on the topic. We also want to thank Springer for implementation of this project as well as Thomas Scheper and Ingrid Samide for suggestions, ideas, and patience during the preparation of the volume.

Spring, 2011

Susann Müller
Thomas Bley

Contents

Adv Biochem Engin/Biotechnol (2011) 124: 1–19
DOI: 10.1007/10_2010_81

Published Online: 12 November 2010

Light Microscopic Analysis of Mitochondrial Heterogeneity in Cell Populations and Within Single Cells

Stefan Jakobs, Stefan Stoldt and Daniel Neumann

Abstract Heterogeneity in the shapes of individual multicellular organisms is a daily experience. Likewise, even a quick glance through the ocular of a light microscope reveals the morphological heterogeneities in genetically identical cultured cells, whereas heterogeneities on the level of the organelles are much less obvious. This short review focuses on intracellular heterogeneities at the example of the mitochondria and their analysis by fluorescence microscopy. The overall mitochondrial shape as well as mitochondrial dynamics can be studied by classical (fluorescence) light microscopy. However, with an organelle diameter generally close to the resolution limit of light, the heterogeneities within mitochondria cannot be resolved with conventional light microscopy. Therefore, we briefly discuss here the potential of subdiffraction light microscopy (nanoscopy) to study inner-mitochondrial heterogeneities.

Keywords Fluorescence microscopy · Mitochondria · Nanoscopy · Single-cell heterogeneity · Super-resolution microscopy

Abbreviations

GSD microscopy	Ground state depletion microscopy
GSDIM	Ground state depletion microscopy followed by individual molecule return
MMP	Mitochondrial membrane potential
PALM	Photoactivated localization microscopy
RESOLFT	Reversible saturable/switchable optical linear (fluorescence) transitions
STED microscopy	Stimulated emission depletion microscopy
STORM	Stochastic optical reconstruction microscopy
TOM	Translocase of the outer membrane

S. Jakobs (✉), S. Stoldt and D. Neumann
Mitochondrial Structure and Dynamics Group,
Max Planck Institute for Biophysical Chemistry, Am Fassberg 11,
37077 Goettingen, Germany
e-mail: sjakobs@gwdg.de

Contents

1 Mitochondria: The Powerhouses of the Cell

Oxidative phosphorylation (OXPHOS) takes place in mitochondria, and thus these organelles are crucial for the regeneration of ATP from ADP and inorganic phosphate in respiration [112]. They are the 'powerhouses of the cell' [84, 115]. In addition, several other essential metabolic pathways take place in these organelles, such as the β-oxidation of fatty acids [66], the formation of iron-sulfur centers [76], the urea cycle, as well as the biogenesis of pyridines, nucleotides and phospholipids [116]. Mitochondria take part in the cellular Ca^{2+} homeostasis [61, 127], and they play an important role during the progression of programmed cell death referred to as apoptosis [25, 42, 70].

Especially in the last 2 decades, it has become clear that mitochondria also play a crucial role in a number of human diseases [36, 94], including diabetes mellitus [93], cancer [12, 15, 86], neurodegenerative diseases such as Parkinson's and Alzheimer's [72, 106, 128], and several others. Further, mitochondrial (dis-) function has been linked to the cellular aging processes, characterized by impaired levels of oxidative phosphorylation and increasing amounts of reactive oxygen species [88, 115].

2 Mitochondrial Heterogeneity Between Different Species

The term mitochondrium [6] is derived from the Greek words *mitos*, which stands for fiber or thread, and *chondros*, which means grain or corn. Put together, they can be translated as a thread-like grain; thus, the word already indicates the heterogeneity of the mitochondrial morphology, which has been known since this organelle was first described [68].

In the last 3 decades, most studies investigating the morphology and dynamics of mitochondria have relied on various forms of far-field fluorescence microscopy imaging mitochondria that were stained with specific fluorescent dyes or tagged with fluorescent proteins [62]. Mitochondrial morphology and dynamics have been studied in many eukaryotic organisms, ranging from monocellular yeasts [63, 95, 96] to higher multicellular eukaryotes, including plants [3, 77, 122] and mammals [7, 27, 71, 75, 108].

Examples of diverse mitochondrial shapes in different organisms are shown in Fig. 1. The spherical mitochondria in the guard cells of a tobacco plant (Fig. 1b) vary considerably from the tubular mitochondria in budding yeast (Fig. 1a) or from the complex mitochondrial networks of a cultivated human cancer cell (Fig. 1c). Mitochondria in plant cells often do not form a continuous network, and they are frequently located next to chloroplasts. Indeed, plant mitochondria differ substantially from mitochondria of other eukaryotes in a number of aspects, including different strategies for genome maintenance, genetic decoding, gene regulation and organelle segregation [9, 80, 104, 105]. In the plant Arabidopsis thaliana, the mitochondrial proteome encompasses about 3,000 proteins [10], whereas about 1,000 different mitochondrial proteins were predicted for budding yeast [85, 103] and about 1,500 for human cells [85].

3 Mitochondrial Heterogeneity Between Different Cell Types

In most cultivated mammalian cells mitochondria form a complex, more or less connected network [39]. These mitochondria have a similar set of proteins and fulfill similar metabolic needs; still, their actual shape may vary considerably:

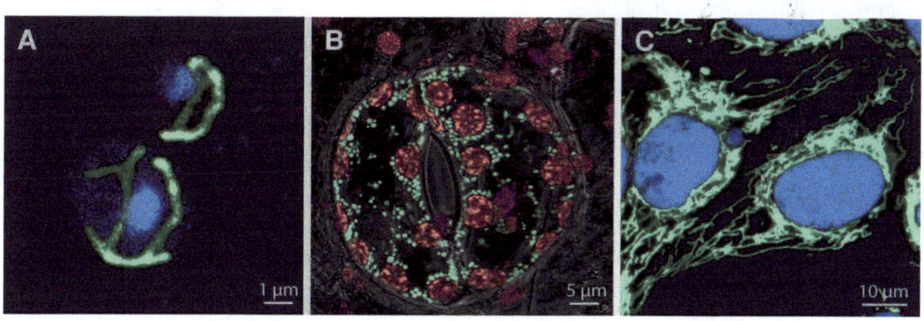

Fig. 1 Different mitochondrial morphologies in fungi, plant and mammalian cells. Fluorescence micrographs of the budding yeast *Saccharomyces cerevisiae* (**a**), a guard cell of the tobacco plant *Nicotiana tabacum* (**b**) and a cultured human (U2OS) cell (**c**). In **a** and **b** the mitochondria (*green*) were labeled by the expression of the green fluorescent protein (GFP) targeted to the mitochondrial matrix. In **c** the mitochondria (*green*) were immunostained by an antibody against the mitochondrial outer membrane protein Tom20. The chloroplasts (**b**) (*red*) were visualized using their strong autofluorescence. The nuclei (**a, c**) (*blue*) were highlighted with DAPI

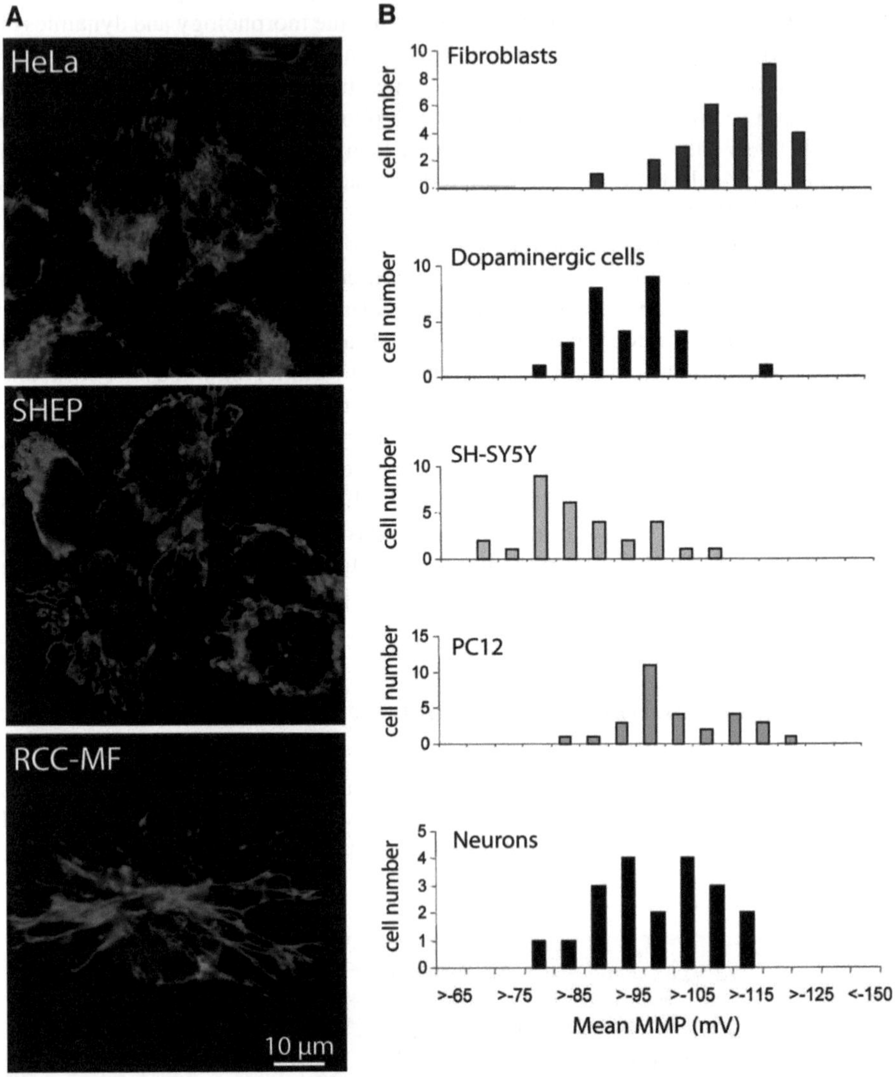

Fig. 2 Mitochondrial heterogeneities between different mammalian cell types. **a** The overall mitochondrial morphologies differ between three human carcinoma cell types (HeLa cervical carcinoma, SHEP neuroblastoma, RCC-MF renal cell carinoma). Mitochondria were immunostained with an antiserum against the mitochondrial protein Tom20; the nuclei were labeled with DAPI. Shown are representative images taken with an epi-fluorescence microscope. **b** The mitochondrial membrane potential (MMP) reflects the functional status of mitochondria. The mean MMP varies significantly between different mammalian cell types (from [58], with permission)

Whereas the mitochondria of HeLa (derived from a human cervical carcinoma) and SHEP (derived from a human neuroblastoma) cells build up a highly interconnected, dense meshwork of slightly curled mitochondrial tubules; the mitochondria of RCC-MF (human renal cell carcinoma) cells are much less interconnected, and the straight mitochondrial tubules appear to radiate from the nucleus (Fig. 2a). However, a more spherical grain-like morphology has been observed in a few cell types [2, 20]. Moreover, in addition to these overall shape differences, also more subtle but characteristic distinctions in the diameter of the mitochondrial tubules of different cell types have been reported [31].

Morphologic shape differences in genetically similar cells are not restricted to closely related mammalian cell lines, but have also been described in other kingdoms. For example, in most plant tissues, the mitochondria exhibit spherical structures of uniform diameter. However, in certain cell types within the vascular tissue, the shapes range from sausage-shaped to long worm-like forms [78].

In addition to these morphological variations, Huang et al. reported on different mitochondrial membrane potentials in various mammalian cell types (Fig. 2b). The mitochondrial membrane potential is a key indicator of cellular viability, as it reflects the pumping of protons across the inner membrane during the process of electron transport and oxidative phosphorylation. The mean MMP in the analyzed cell types ranged from -112 ± 2 mV in fibroblasts to -87 ± 2 mV in SH-SY5Y cells, suggesting specific adaptations to the energy demands of the respective cells [58].

Hence, there can be large differences in the shapes of the mitochondria of different cells. Some of these differences may be attributed to different functional tasks of the respective cells, whereas other shapes are less obvious to comprehend. Obviously, cells have a wide repertoire of potential mitochondrial morphologies that may be adapted to the specific metabolic state of the cell.

4 Mitochondrial Shape Changes in a Cell Over Time

4.1 Structural Adaptations to the Cellular Energy Demands

The main physiological role of mitochondria is to generate ATP. Frequently, a connection between mitochondrial structure and the bioenergetical requirements of the cell has been suggested [5, 13, 109, 131]. For example, in budding yeast cells (*S. cerevisiae*), the volume of the mitochondrial reticulum is increased up to three-fold after changing from a fermentable (glucose) to a non-fermentable (glycerol) carbon source (Fig. 3a) [34, 125]. When glucose is available in large amounts in the budding yeast, respiration is repressed and ATP is primarily produced via glycolysis (without involvement of mitochondria), a phenomenon also known as the "Crabtree effect" [24, 26, 29, 37]. Therefore, only under non-fermentable

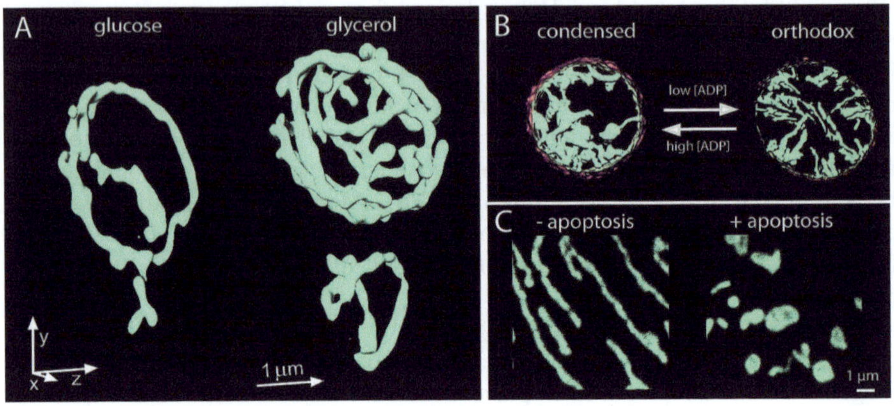

Fig. 3 Mitochondrial shape adaptations to different conditions. **a** The mitochondrial network of the budding yeast *S. cerevisiae* adapts to different carbon sources in the growth medium. Shown are 3D reconstructions of the mitochondria of living cells labeled with the green fluorescent protein (GFP). The cells were grown in a fermentable (glucose) or a non-fermentable (glycerol) growth medium. Images were taken with a multifocal multiphoton 4Pi-confocal microscope (from [34], with permission). **b** Electron tomographic reconstructions of rat liver mitochondria reveal changes in the mitochondrial inner membrane topology associated with the orthodox-condensed transition. The mitochondrial outer membrane is shown in *red*, the inner boundary membrane in *yellow* and the cristae in *green*. The depicted mitochondria have diameters of 1,500 nm (*left*) and 500 nm (*right*) (from [82], with permission). **c** Mitochondria in unchallenged cultivated mammalian cells (U2OS) adopt a tubular shape, forming an interconnected network (*left*). Upon induction of apoptosis (10 µM actinomycin D for 12 h), the network fragments, resulting in numerous small mitochondria with a spherical shape (*right*). For visualization, the mitochondria were labeled with an antiserum against the mitochondrial protein Tom20 and imaged with a confocal microscope

conditions are mitochondria required for ATP production. This is a clear example of mitochondrial shape adaptations to different energy needs.

Likewise, it has been reported that in drought-stressed spinach leaves, a decrease of the mitochondrial volume in parenchyma cells can be observed [138]. This reduced mitochondrial volume has been suggested to be caused by glucose starvation deriving from decreased photosynthetic activity.

The energy requirements of the cell may also influence the inner structure of the mitochondria. Depending on the ADP concentration, the architecture of the inner membrane can change between a condensed and an orthodox state [43]. Electron tomography revealed that in the condensed state, the matrix is compacted, and the cristae form large compartments with multiple tubular connections to the peripheral region as well as to each other. In the orthodox state, the matrix is expanded, and the cristae tend to be tubes or short flat lamellae with one or two openings in the peripheral region of the inner membrane (Fig. 3b) [82]. It has been suggested that this morphological transition could result in the elimination of diffusion bottlenecks inside large intracristal compartments that would otherwise reduce the efficiency of ATP production [82, 83].

4.2 Fusion and Fission

In healthy cells, mitochondrial fusion and fission events are in an equilibrium so that their relative rates determine the average size of the individual mitochondrial tubules and the degree of network connectivity [7, 18, 19, 56, 63, 95]. In the fission yeast *Schizosaccharomyces pombe*, the equilibrium is shifted during mitosis to fission, resulting in highly fragmented mitochondria [65]. A similar observation has been made in budding yeast cells undergoing meiosis [41]. A tempting explanation for these temporal fragmentations is the conversion of a low copy number organelle into a high copy number one, thus increasing the chance to distribute a sufficient number of mitochondria for each daughter cell to commence the next cell cycle [65, 79, 130].

Another process where excessive fission takes place is apoptosis. Upon induction of the cell death program, the mitochondrial network disintegrates, yielding numerous and smaller mitochondria (Fig. 3c) [17, 64, 137]. Whether this fragmentation of the mitochondrial network has a functional relevance for the ongoing cell death program or if it is just a by-product is still under debate [16, 38, 73, 98, 123, 137]. Apoptosis is also accompanied by a remodeling of the mitochondrial inner membrane [120, 129, 135, 136].

5 Mitochondrial Heterogeneity Within a Single Cell at a Certain Time

5.1 Morphological Heterogeneity

The distribution and shape of the mitochondria in many cultured mammalian cells are rather heterogeneous. Although they may disperse throughout the whole cytosol, frequently a pronounced tendency of aggregation around the nucleus can be observed (Figs. 1, 2a) [7, 20, 39]. But are the mitochondria of a cell luminally continuous? To address this question, Collins and coworkers performed FRAP (fluorescence recovery after photobleaching) experiments with different cell types including HeLa, PAEC, COS-7, HUVEC, cortical astrocytes and neuronal cells expressing a fluorescent protein in the mitochondrial matrix [20]. It was observed that the fluorescence in the bleached regions did not recover to >10% of its initial value 1 h after irradiation. These data suggest that in the analyzed cells, the mitochondria were largely disconnected, possibly indicating functional heterogeneities of the mitochondria within a single cell.

5.2 Functional Heterogeneity

For pancreatic acinar cells, the existence of three distinct groups of mitochondria with different functions was shown [97]: perigranular mitochondria, perinuclear

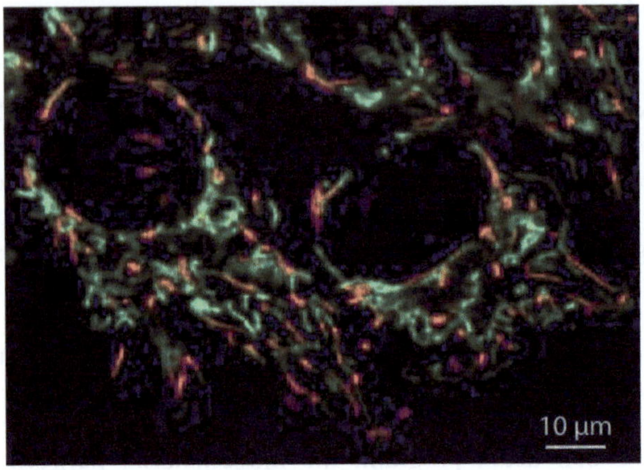

Fig. 4 Heterogeneity in the functional status of mitochondria within a single cell. Heterogeneity in the mitochondrial membrane potential is revealed by the dye JC-1 in HeLa cells. JC-1 is a green fluorescent monomer at low membrane potentials and forms orange/red fluorescent aggregates at high membrane potentials. Thus, the different colors of the fluorescence emission indicate the heterogeneity of the mitochondrial membrane potential in a single cell (from [20])

mitochondria and peripheral mitochondria near the basal plasma membrane. Photobleaching experiments indicated that these three groups are not luminally connected and respond differently to cytosolic Ca^{2+} signals. Therefore, it was suggested that they participate in the local regulation of Ca^{2+} homeostasis.

Differences between mitochondria within a single cell were not only revealed by their response to Ca^{2+} levels, but also by their membrane potential. The membrane potential is a potent functional readout of mitochondrial activity and can be monitored by using positively charged, lipophilic fluorescent dyes [62, 74, 92, 113], including JC-1 (tetrachloro-1,1,3,3-tetraethylbenzimidazol-carbocyanine-iodide) [124] and TMRM (tetramethylrhodamine-methyl-ester) [35]. JC-1 is a green fluorescent monomer at low membrane potentials and forms orange/red fluorescent aggregates at high membrane potentials. The appropriateness of JC-1 has been controversially discussed, because it provides only a qualitative readout on the membrane potential, and the staining efficiency is concentration and salt dependent [92, 126, 133]. Still, using JC-1, it was demonstrated that within a single HeLa cell, mitochondria exhibiting green fluorescence coexist with red fluorescing mitochondria (Fig. 4) [20]. Heterogeneities in the membrane potential were also conclusively reported using TMRM allowing quantitative studies [14, 28, 30, 126, 132].

These findings demonstrate that even in cells with morphologically rather similar mitochondria, pools of functionally distinct mitochondria may exist. Hence, presumably not unexpectedly, functional heterogeneities within the mitochondria were also observed in nerve cells, which exhibit a pronounced spatial and functional asymmetry. For example, in one of the largest nerve terminals in the central nervous system of mammals, the calyx of Held [45], two mitochondrial subpopulations were described: a small mitochondrial population with complex geometries located near the presynaptic membrane called the mitochondria-associated adherens complex (MAC) and a large mitochondrial pool with a simpler architecture, which was not preferentially located near presynaptic membranes [110]. The MAC-forming mitochondria were suggested to play a central role in high rate, temporally precise neurotransmission. Further, electron tomography revealed that these MAC-forming mitochondria have a specialized ultrastructure exhibiting a polarized cristae architecture in that cristae junctions were aligned with the cytoskeleton and occurred at higher density in the mitochondrial membrane that faces the presynaptic membrane [102].

These data conclusively indicate that, at least in some cell types, morphologically and functionally distinct mitochondrial subpopulations exist. However, very little is known about whether such functional differences are due to different protein distributions within the mitochondria.

6 Nanoscopy of Protein Distributions in Mitochondria

Almost all that we know about the inner architecture of mitochondria comes from electron microscopy and electron tomography data. These techniques are very powerful for dissecting the membrane architecture of the organelles, but are generally less well suited to study the distributions of proteins, requiring specific labeling, ideally in living cells. For these challenges, fluorescence microscopy is generally the method of choice.

However, the wave nature of light imposes a seemingly fundamental limit to the attainable resolution of light microscopes. According to Abbe, the resolution limitation is ultimately rooted in the phenomenon of diffraction [1]. Because of diffraction, focusing of light always results in a blurred spot [11], whose size determines the resolution. Thus, the highest achievable resolution with objective lenses and visible light is $\sim$180 nm in the imaging plane. When using a single lens, the resolution along the optical axis is inescapably worse; even confocal or two photon fluorescence microscopes, which stand out in their ability to provide 3D images by optical sectioning, can only distinguish fluorescent objects if their axial separation is at least 500–800 nm.

For this reason, it has been impossible to resolve protein distributions within the inner membrane of unaltered mitochondria using light microscopy [134]. Likewise, the protein density in mitochondria is apparently so high that also protein complexes in the mitochondrial matrix or in the outer membrane have proved to be non-resolvable by conventional light microscopy. In fact, not very long ago, obtaining a spatial resolution sufficient to resolve inner-mitochondrial features with an optical microscope that uses lenses and focused visible light was considered unfeasible.

6.1 Concepts to Overcome the Diffraction Barrier

In recent years, a number of 'nanoscopy' or 'superresolution' fluorescence microscopy techniques have been invented to fundamentally overcome the diffraction barrier. A number of excellent and exhaustive reviews describing the theory as well as the practical details are available [47, 48, 57, 99]. Therefore, we give here only a brief overview of the concepts of the various nanoscopy schemes.

Stimulated emission depletion (STED) microscopy [53] and ground state depletion (GSD) microscopy [52] were the first concrete and viable physical concepts to fundamentally overcome the limiting role of diffraction in a lens-based optical microscope. In brief, STED and GSD use a selected pair of bright (fluorescent) and dark (non-fluorescent) fluorophore states to restrict the bright state to subdiffraction dimensions. To this end, optical transitions are utilized that allow one to transiently switch off the ability of the dye to fluoresce by confining the dye to a dark state. The transition is effected with a light intensity distribution featuring a zero, switching the fluorescence off everywhere except at the zero where the fluorophore is still allowed to be bright. Moving the zero across the specimen switches the signal of adjacent features sequentially on and off, allowing their separate registration. This allows one to image fine structures that would otherwise be blurred in a conventional diffraction-limited image. Hence, the spatial confinement of molecular states allows the elimination of the resolution-limiting effect of diffraction without eliminating the diffraction.

In its initial demonstration, STED microscopy was realized as a point-scanning system (Fig. 5), whereby the excitation focus was a normal confocal spot and the

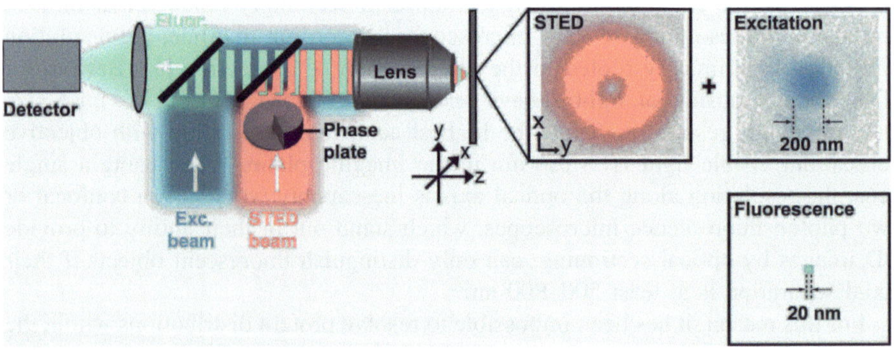

Fig. 5 Schematic drawing of a point-scanning STED microscope. Excitation and STED are accomplished with synchronized laser pulses focused by a lens onto the sample, sketched as *blue* and *red* beams, respectively. Fluorescence is registered with a detector. A phase plate is placed in the light path of the STED beam to create a ring-shaped focus featuring an intensity zero in its center. Measured intensity distributions in the focus are shown on the *right*. The diffraction limited excitation focus is overlapped with the ring-shaped STED focus. Saturated depletion confines the region of excited molecules to the zero, leaving an effective focus of subdiffraction dimensions

STED focus resembled a doughnut, featuring a light intensity zero in the center [67]. Figure 5 shows a typical experimental focal intensity distribution of the excitation spot (blue), overlapped with a STED-spot (red) featuring a central intensity zero. Saturated depletion inhibits the fluorescence everywhere except at the very center of the focal region.

Later, the concept of STED and GSD microscopy was expanded to photoswitching molecules, including synthetic organic molecules and reversibly photoswitchable fluorescent proteins (RSFPs). This family of approaches was named RESOLFT, standing for reversible saturable/switchable optical linear (fluorescence) transitions [46, 50, 51]. The RESOLFT concepts are purely "physical" or "physicochemical" concepts, because the subdiffraction resolution is a direct consequence of the molecular transition employed. Because the position of the zero is defined with the RESOLFT concepts, e.g., it is defined where the molecules are "on" and where they are "off," these concepts operate with any number of molecules, ranging from single to many (Fig. 6a).

The concept of switching is also essential in more recent far-field fluorescence nanoscopy approaches referred to here as superresolution microscopy by single-molecule switching and localization, which differ from the RESOLFT concepts by the fact that they switch molecules stochastically in space and utilize mathematics to assemble the image (Fig. 6). This family of approaches has been initially implemented independently by several groups and named photoactivated localization microscopy (PALM) [8], fluorescence photoactivated localization microscopy (FPALM) [55] and stochastic optical reconstruction microscopy (STORM) [111].

The operating principle of these concepts is to start with the vast majority of labels in an inactive (dark) state, not contributing to the fluorescence (Fig. 6b). A small fraction ($\ll 1\%$) is then stochastically transferred to the fluorescence state so that the single molecules can be individually imaged and localized to give nanometer-level precision coordinates. After the coordinates have been recorded, the bright fluorophores are then removed (e.g., by bleaching, thermal relaxation, or otherwise) so that a new subset of the fluorophores can be transferred into the fluorescent state and recorded to obtain an additional set of molecular coordinates. This process is repeated thousands of times until a sufficient number of molecular coordinates is recorded. Importantly, the molecular coordinates do not have infinite localization accuracy, but the localization accuracy depends on the number of emitted photons from the individually localized single molecule. Finally, a composite single-molecule nanoscopy image of all these coordinates is created.

6.2 Nanoscopy on Mitochondria

As described above, several studies using conventional light microscopy or electron microscopy have provided ample evidence on morphological and functional heterogeneities of mitochondria within single cells. However, due to the previous

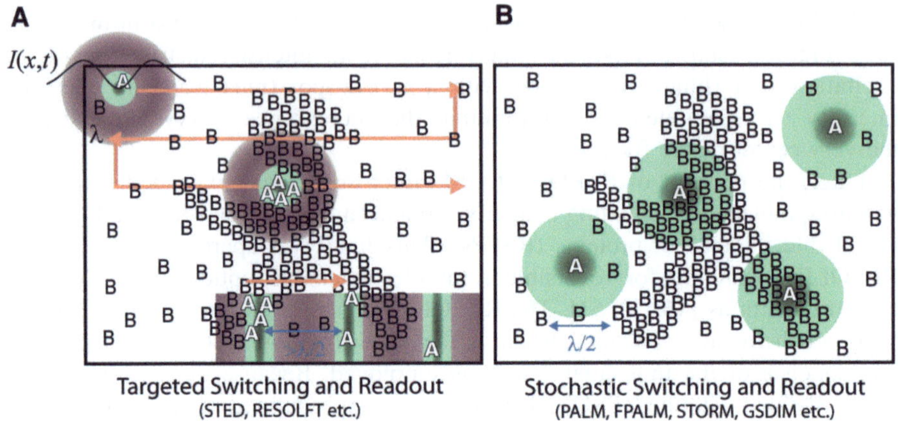

Fig. 6 Concepts to overcome the diffraction barrier. To resolve image details that are closer than the diffraction limit, all far-field fluorescence nanoscopy concepts realized so far switch the fluorophore between two distinguishable states, a bright state *A* and a dark state *B*, to construct subdiffraction images. **a** In the targeted readout mode, a spatial light intensity distribution *I*(*x*, *t*) having a zero intensity point in space switches the molecules such that one of the states (here: *A*) is confined to subdiffraction dimensions. The image is assembled by scanning the zero over the sample and recording adjacent features sequentially in time. To parallelize the recording procedure, the zero can also be *line shaped* or an array of *zero lines*. As the zero is translated across the object, the molecules undergo several times the transition $B \rightarrow A \rightarrow B$, which explains the need for reversible transitions $A \longleftrightarrow B$. **b** The stochastic readout mode detects single fluorophores from a random position within the diffraction zone. To this end, a molecule is transferred to a state *A* that is able to emit $m \gg 1$ photons in a row, while the neighboring molecules remain in the dark state *B*. The distance between molecules in state *A* should be larger than the diffraction limit. The detection of $m \gg 1$ photons allows the calculation of the coordinate of emission from the centroid of the diffraction fluorescence spot formed on a camera. After the recording, the molecule is switched off to *B* in order to allow the recording of an adjacent molecule. If it is sufficient to record a single picture, the stochastic readout requires each molecule to cycle only once $B \rightarrow A \rightarrow B$ (adapted from [49], with permission)

lack of appropriate technologies, it is still largely unclear whether these heterogeneities also reflect heterogeneities in the sub-mitochondrial distribution of proteins. Until the advent of the nanoscopy concepts detailed above, it had proved to be very challenging or even impossible to address such questions because many mitochondrial protein complexes could not be resolved with conventional microscopy due to the limited resolution. For example, the TOM complexes, which are the primary import pores for nuclear encoded mitochondrial proteins, proved to be so densely packed in the mitochondrial outer membrane that it required STED microscopy to reveal individual TOM clusters in the mitochondrial outer membrane (Fig. 7a) [32]. Likewise, STED microscopy in combination with co-localization algorithms was used to determine quantitatively the degree of co-localization between hexokinase-I and each of the three isoforms of the human voltage-dependent anion-selective channel (hVDAC). The nanoscopy data showed

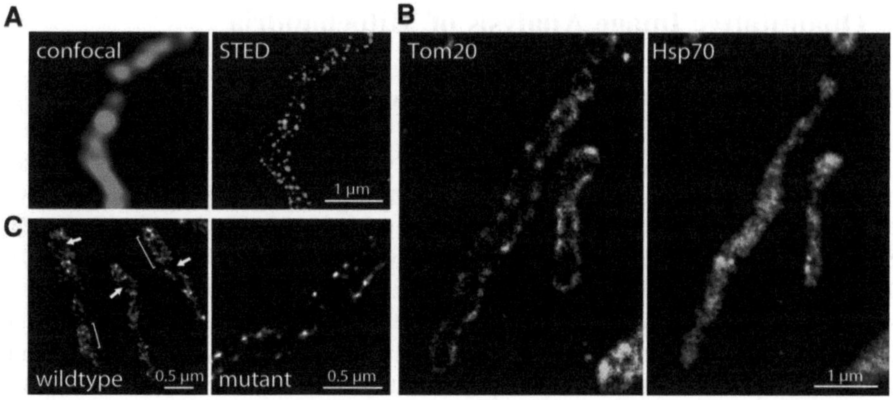

Fig. 7 STED microscopy reveals sub-mitochondrial protein distributions on the nanoscale. **a** Comparison of images taken with a (diffraction-limited) confocal microscope (*left*) and a STED microscope (*right*). The mitochondria of a PtK2 (rat kangaroo kidney) cell were labeled with an antibody against the outer membrane protein Tom20. In case of the STED image, individual Tom20 clusters are resolved, whereas they are blurred in the confocal case. **b** Two-color isoSTED images of mitochondria in Vero (African green monkey) cells allow distinguishing between proteins localized in the outer mitochondrial membrane (Tom20, *left*) and proteins of the mitochondrial matrix (Hsp70, *right*). **c** With the isoSTED approach, it is possible to reveal the arrangement of cristae, here in mammalian PtK2 cells (*left*). The cells were labeled with antibodies against the inner membrane protein complex F_1F_0ATP-synthase. The brackets indicate parallel cristae arrangements, and the arrows indicate regions devoid of cristae. Alterations in the cristae structure could be observed when depleting the protein mitofilin (*right*), which controls cristae morphology (modified from [117] and [118], with permission)

that the degree of co-localization between hVDAC and hexokinase-I is isoform-specific, suggesting a more complex interplay of these proteins than previously anticipated [91].

Recently, the implementation of STED with opposing lenses, called isoSTED microscopy, has enabled the recording of the interior of mitochondria with a 3D resolution of better than 50 nm in all room directions [117]. Using this approach, Schmidt et al. analyzed the distributions of various proteins within these organelles (Fig. 7b) [117]. By labeling the F_1F_0ATPase, a protein complex that lines the inner membrane, these authors could also delineate the flow of the inner membrane in the mitochondria of intact cells, revealing heterogeneities in the cristae arrangements (Fig. 7c) [118].

These data conclusively demonstrate that with the now widely available nanoscopy approaches, it is possible to analyze sub-mitochondrial protein distributions. We expect that in the near future these approaches will be utilized to correlate functional heterogeneities with the nanoscale distribution of mitochondrial proteins, providing new insights into the heterogeneity of mitochondria on an additional level.

7 Quantitative Image Analysis of Mitochondria

The pronounced heterogeneity of mitochondria on the functional and morphological level not only between individual cells, but also within them, immediately prohibits a biologically meaningful analysis of subtle mitochondrial differences based on individual images. A plethora of computational tools has been developed to analyze large image data sets quantitatively [4, 21–23, 30, 54, 59, 60, 69, 81, 89–91, 100, 114, 119] and have been reviewed expertly [33, 40, 44, 87, 101, 107, 121]. The availability of nanoscopy/superresolution techniques to a growing scientific community will undoubtedly further increase the need for quantitative data analysis and at the same time will elicit new challenges, including among others the fact that with these techniques the attainable optical resolution is similar to the size of the used fluorescent labels.

Acknowledgments We thank C.A. Wurm for insightful discussions and providing some of the STED images. We also thank R. Schmidt and A. Egner regarding STED microscopy of mitochondria, S.W. Hell for continuous support and J. Jethwa for carefully reading the manuscript. Part of the work reported in this review was supported by the Bundesministerium für Bildung und Forschung (BMBF) (SysCompart, to S.J.).

References

1. Abbe E (1873) Beiträge zur Theorie des Mikroskops und der mikroskopischen Wahrnehmung. Arch f Mikroskop Anat 9:413–420
2. Amchenkova AA, Bakeeva LE, Chentsov YS et al (1988) Coupling membranes as energy-transmitting cables. I. Filamentous mitochondria in fibroblasts and mitochondrial clusters in cardiomyocytes. J Cell Biol 107:481–495
3. Arimura S, Tsutsumi N (2002) A dynamin-like protein (ADL2b), rather than FtsZ, is involved in *Arabidopsis* mitochondrial division. Proc Natl Acad Sci USA 99:5727–5731
4. Barbe L, Lundberg E, Oksvold P et al (2008) Toward a confocal subcellular atlas of the human proteome. Mol Cell Proteomics 7:499–508
5. Benard G, Bellance N, James D et al (2007) Mitochondrial bioenergetics and structural network organization. J Cell Sci 120:838–848
6. Benda C (1898) Ueber dier Spermatogenese de Verbebraten und höherer Evertebraten, II. Theil: Die Histogenese der Spermien. Arch Anat Physiol 73:393–398
7. Bereiter-Hahn J (1990) Behavior of mitochondria in the living cell. Int Rev Cytol 122:1–63
8. Betzig E, Patterson GH, Sougrat R et al (2006) Imaging intracellular fluorescent proteins at nanometer resolution. Science 313:1642–1645
9. Bodenstein-Lang J, Buch A, Follmann H (1989) Animal and plant mitochondria contain specific thioredoxins. FEBS Lett 258:22–26
10. Bogorad L (2008) Evolution of early eukaryotic cells: genomes, proteomes, and compartments. Photosynth Res 95:11–21
11. Born M, Wolf E (2002) Principles of optics. Cambridge University Press, Cambridge
12. Brandon M, Baldi P, Wallace DC (2006) Mitochondrial mutations in cancer. Oncogene 25:4647–4662
13. Brocard JB, Rintoul GL, Reynolds IJ (2003) New perspectives on mitochondrial morphology in cell function. Biol Cell 95:239–242

14. Buckman JF, Reynolds IJ (2001) Spontaneous changes in mitochondrial membrane potential in cultured neurons. J Neurosci 21:5054–5065
15. Carew JS, Huang P (2002) Mitochondrial defects in cancer. Mol Cancer 1:9
16. Cassidy-Stone A, Chipuk JE, Ingerman E et al (2008) Chemical inhibition of the mitochondrial division dynamin reveals its role in Bax/Bak-dependent mitochondrial outer membrane permeabilization. Dev Cell 14:193–204
17. Cereghetti GM, Scorrano L (2006) The many shapes of mitochondrial death. Oncogene 25:4717–4724
18. Cerveny KL, Tamura Y, Zhang Z et al (2007) Regulation of mitochondrial fusion and division. Trends Cell Biol 17:563–569
19. Chan DC (2006) Mitochondrial fusion and fission in mammals. Annu Rev Cell Dev Biol 22:79–99
20. Collins TJ, Berridge MJ, Lipp P et al (2002) Mitochondria are morphologically and functionally heterogeneous within cells. EMBO J 21:1616–1627
21. Comeau JW, Costantino S, Wiseman PW (2006) A guide to accurate fluorescence microscopy colocalization measurements. Biophys J 91:4611–4622
22. Conrad C, Gerlich DW (2010) Automated microscopy for high-content RNAi screening. J Cell Biol 188:453–461
23. Costes SV, Daelemans D, Cho EH et al (2004) Automatic and quantitative measurement of protein-protein colocalization in live cells. Biophys J 86:3993–4003
24. Crabtree HG (1929) Observations on the carbohydrate metabolism of tumours. Biochem J 23:536–545
25. Danial NN, Korsmeyer SJ (2004) Cell death: critical control points. Cell 116:205–219
26. De Deken RH (1966) The Crabtree effect: a regulatory system in yeast. J Gen Microbiol 44:149–156
27. Detmer SA, Chan DC (2007) Functions and dysfunctions of mitochondrial dynamics. Nat Rev Mol Cell Biol 8:870–879
28. Diaz G, Falchi AM, Gremo F et al (2000) Homogeneous longitudinal profiles and synchronous fluctuations of mitochondrial transmembrane potential. FEBS Lett 475:218–224
29. Diaz-Ruiz R, Averet N, Araiza D et al (2008) Mitochondrial oxidative phosphorylation is regulated by fructose 1,6-bisphosphate. A possible role in Crabtree effect induction? J Biol Chem 283:26948–26955
30. Distelmaier F, Koopman WJ, Testa ER et al (2008) Life cell quantification of mitochondrial membrane potential at the single organelle level. Cytometry A 73:129–138
31. Dlaskova A, Spacek T, Santorova J et al (2010) 4Pi microscopy reveals an impaired three-dimensional mitochondrial network of pancreatic islet beta-cells, an experimental model of type-2 diabetes. Biochim Biophys Acta 1797:1327–1341
32. Donnert G, Keller J, Wurm CA et al (2007) Two-color far-field fluorescence nanoscopy. Biophys J 92:L67–69
33. Dragunow M (2008) High-content analysis in neuroscience. Nat Rev Neurosci 9:779–788
34. Egner A, Jakobs S, Hell SW (2002) Fast 100-nm resolution three-dimensional microscope reveals structural plasticity of mitochondria in live yeast. Proc Natl Acad Sci USA 99: 3370–3375
35. Ehrenberg B, Montana V, Wei MD et al (1988) Membrane potential can be determined in individual cells from the nernstian distribution of cationic dyes. Biophys J 53:785–794
36. Fernandez-Moreno MA, Bornstein B, Petit N et al (2000) The pathophysiology of mitochondrial biogenesis: towards four decades of mitochondrial DNA research. Mol Genet Metab 71:481–495
37. Fiechter A, Fuhrmann GF, Kappeli O (1981) Regulation of glucose metabolism in growing yeast cells. Adv Microb Physiol 22:123–183
38. Frank S, Gaume B, Bergmann-Leitner ES et al (2001) The role of dynamin-related protein 1, a mediator of mitochondrial fission, in apoptosis. Dev Cell 1:515–525
39. Frazier AE, Kiu C, Stojanovski D et al (2006) Mitochondrial morphology and distribution in mammalian cells. Biol Chem 387:1551–1558

40. Glory E, Murphy RF (2007) Automated subcellular location determination and high-throughput microscopy. Dev Cell 12:7–16
41. Gorsich SW, Shaw JM (2004) Importance of mitochondrial dynamics during meiosis and sporulation. Mol Biol Cell 15:4369–4381
42. Green DR, Kroemer G (2004) The pathophysiology of mitochondrial cell death. Science 305:626–629
43. Hackenbrock CR (1966) Ultrastructural bases for metabolically linked mechanical activity in mitochondria. I. Reversible ultrastructural changes with change in metabolic steady state in isolated liver mitochondria. J Cell Biol 30:269–297
44. Hamilton N (2009) Quantification and its applications in fluorescent microscopy imaging. Traffic 10:951–961
45. Held H (1893) Die centrale Gehorleitung. Arch f Anat u Physiol Anat Abt 201–248
46. Hell SW (2003) Toward fluorescence nanoscopy. Nat Biotechnol 21:1347–1355
47. Hell SW (2007) Far-field optical nanoscopy. Science 316:1153–1158
48. Hell SW (2009a) Far-field optical nanoscopy. In: Single molecule spectroscopy in chemistry, physics and biology. Springer, Berlin
49. Hell SW (2009b) Microscopy and its focal switch. Nat Methods 6:24–32
50. Hell SW, Dyba M, Jakobs S (2004) Concepts for nanoscale resolution in fluorescence microscopy. Curr Opin Neurobiol 14:599–609
51. Hell SW, Jakobs S, Kastrup L (2003) Imaging and writing at the nanoscale with focused visible light through saturable optical transitions. Appl Phys A (Materials Science Processing) 77:859–860
52. Hell SW, Kroug M (1995) Ground-state depletion fluorescence microscopy, a concept for breaking the diffraction resolution limit. Appl Phys B 60:495–497
53. Hell SW, Wichmann J (1994) Breaking the diffraction resolution limit by stimulated emission: stimulated emission depletion microscopy. Opt Lett 19:780–782
54. Herold J, Schubert W, Nattkemper TW (2010) Automated detection and quantification of fluorescently labeled synapses in murine brain tissue sections for high throughput applications. J Biotechnol. doi:10.1016/j.jbiotec.2010.03.004
55. Hess ST, Girirajan TP, Mason MD (2006) Ultra-high resolution imaging by fluorescence photoactivation localization microscopy. Biophys J 91:4258–4272
56. Hoppins S, Lackner L, Nunnari J (2007) The machines that divide and fuse mitochondria. Annu Rev Biochem 76:751–780
57. Huang B, Bates M, Zhuang X (2009) Super-resolution fluorescence microscopy. Annu Rev Biochem 78:993–1016
58. Huang HM, Fowler C, Zhang H et al (2004) Mitochondrial heterogeneity within and between different cell types. Neurochem Res 29:651–658
59. Huh WK, Falvo JV, Gerke LC et al (2003) Global analysis of protein localization in budding yeast. Nature 425:686–691
60. Hutchins JR, Toyoda Y, Hegemann B et al (2010) Systematic analysis of human protein complexes identifies chromosome segregation proteins. Science 328:593–599
61. Jacobson J, Duchen MR (2004) Interplay between mitochondria and cellular calcium signalling. Mol Cell Biochem 256–257:209–218
62. Jakobs S (2006) High resolution imaging of live mitochondria. Biochim Biophys Acta 1763:561–575
63. Jakobs S, Martini N, Schauss AC et al (2003) Spatial and temporal dynamics of budding yeast mitochondria lacking the division component Fis1p. J Cell Sci 116:2005–2014
64. Jourdain A, Martinou JC (2009) Mitochondrial outer-membrane permeabilization and remodelling in apoptosis. Int J Biochem Cell Biol 41:1884–1889
65. Jourdain I, Gachet Y, Hyams JS (2009) The dynamin related protein Dnm1 fragments mitochondria in a microtubule-dependent manner during the fission yeast cell cycle. Cell Motil Cytoskeleton 66:509–523
66. Kennedy EP, Lehninger AL (1949) Oxidation of fatty acids and tricarboxylic acid cycle intermediates by isolated rat liver mitochondria. J Biol Chem 179:957–972

67. Klar TA, Jakobs S, Dyba M et al (2000) Fluorescence microscopy with diffraction resolution barrier broken by stimulated emission. Proc Natl Acad Sci USA 97:8206–8210
68. Kölliker A (1857) Einige Bemerkungen über die Endigungen der Hautnerven und den Bau der Muskeln. Zeitschr f wissensch Zool 8:311–325
69. Koopman WJ, Visch HJ, Smeitink JA et al (2006) Simultaneous quantitative measurement and automated analysis of mitochondrial morphology, mass, potential, and motility in living human skin fibroblasts. Cytometry A 69:1–12
70. Kroemer G (2003) Mitochondrial control of apoptosis: an introduction. Biochem Biophys Res Commun 304:433–435
71. Kuznetsov AV, Hermann M, Saks V et al (2009) The cell-type specificity of mitochondrial dynamics. Int J Biochem Cell Biol 41:1928–1939
72. Kwong JQ, Beal MF, Manfredi G (2006) The role of mitochondria in inherited neurodegenerative diseases. J Neurochem 97:1659–1675
73. Lee YJ, Jeong SY, Karbowski M et al (2004) Roles of the mammalian mitochondrial fission and fusion mediators Fis1, Drp1, and Opa1 in apoptosis. Mol Biol Cell 15:5001–5011
74. Lemasters JJ, Ramshesh VK (2007) Imaging of mitochondrial polarization and depolarization with cationic fluorophores. Methods Cell Biol 80:283–295
75. Liesa M, Palacin M, Zorzano A (2009) Mitochondrial dynamics in mammalian health and disease. Physiol Rev 89:799–845
76. Lill R (2009) Function and biogenesis of iron-sulphur proteins. Nature 460:831–838
77. Logan DC (2006) Plant mitochondrial dynamics. Biochim Biophys Acta 1763:430–441
78. Logan DC, Leaver CJ (2000) Mitochondria-targeted GFP highlights the heterogeneity of mitochondrial shape, size and movement within living plant cells. J Exp Bot 51:865–871
79. Lowe M, Barr FA (2007) Inheritance and biogenesis of organelles in the secretory pathway. Nat Rev Mol Cell Biol 8:429–439
80. Mackenzie S, McIntosh L (1999) Higher plant mitochondria. Plant Cell 11:571–586
81. Manders E, Verbeek FJ, Aten JA (1993) Measurement of co-localization of objects in dual-colour confocal images. J Micros 169:375–382
82. Mannella CA (2006) Structure and dynamics of the mitochondrial inner membrane cristae. Biochim Biophys Acta 1763:542–548
83. Mannella CA, Pfeiffer DR, Bradshaw PC et al (2001) Topology of the mitochondrial inner membrane: dynamics and bioenergetic implications. IUBMB Life 52:93–100
84. McBride HM, Neuspiel M, Wasiak S (2006) Mitochondria: more than just a powerhouse. Curr Biol 16:R551–560
85. Meisinger C, Sickmann A, Pfanner N (2008) The mitochondrial proteome: from inventory to function. Cell 134:22–24
86. Modica-Napolitano JS, Kulawiec M, Singh KK (2007) Mitochondria and human cancer. Curr Mol Med 7:121–131
87. Muzzey D, van Oudenaarden A (2009) Quantitative time-lapse fluorescence microscopy in single cells. Annu Rev Cell Dev Biol 25:301–327
88. Navarro A, Boveris A (2007) The mitochondrial energy transduction system and the aging process. Am J Physiol Cell Physiol 292:C670–686
89. Negishi T, Nogami S, Ohya Y (2009) Multidimensional quantification of subcellular morphology of *Saccharomyces cerevisiae* using CalMorph, the high-throughput image-processing program. J Biotechnol 141:109–117
90. Neumann B, Walter T, Heriche JK et al (2010a) Phenotypic profiling of the human genome by time-lapse microscopy reveals cell division genes. Nature 464:721–727
91. Neumann D, Bückers J, Kastrup L et al (2010b) Two-color STED microscopy reveals different degrees of colocalization between hexokinase-I and the three human VDAC isoforms. PMC Biophys 3:4
92. Nicholls DG, Ward MW (2000) Mitochondrial membrane potential and neuronal glutamate excitotoxicity: mortality and millivolts. Trends Neurosci 23:166–174
93. Nishikawa T, Araki E (2007) Impact of mitochondrial ROS production in the pathogenesis of diabetes mellitus and its complications. Antioxid Redox Signal 9:343–353

94. Nishino I, Kobayashi O, Goto Y et al (1998) A new congenital muscular dystrophy with mitochondrial structural abnormalities. Muscle Nerve 21:40–47
95. Nunnari J, Marshall WF, Straight A et al (1997) Mitochondrial transmission during mating in *Saccharomyces cerevisiae* is determined by mitochondrial fusion and fission and the intramitochondrial segregation of mitochondrial DNA. Mol Biol Cell 8:1233–1242
96. Okamoto K, Shaw JM (2005) Mitochondrial morphology and dynamics in yeast and multicellular eukaryotes. Annu Rev Genet 39:503–536
97. Park MK, Ashby MC, Erdemli G et al (2001) Perinuclear, perigranular and sub-plasmalemmal mitochondria have distinct functions in the regulation of cellular calcium transport. EMBO J 20:1863–1874
98. Parone PA, James DI, Da Cruz S et al (2006) Inhibiting the mitochondrial fission machinery does not prevent Bax/Bak-dependent apoptosis. Mol Cell Biol 26:7397–7408
99. Patterson G, Davidson M, Manley S et al (2010) Superresolution Imaging using Single-Molecule Localization. Annu Rev Phys Chem 61:346–367
100. Peng T, Bonamy GM, Glory-Afshar E et al (2010) Determining the distribution of probes between different subcellular locations through automated unmixing of subcellular patterns. Proc Natl Acad Sci USA 107:2944–2949
101. Pepperkok R, Ellenberg J (2006) High-throughput fluorescence microscopy for systems biology. Nat Rev Mol Cell Biol 7:690–696
102. Perkins GA, Tjong J, Brown JM et al (2010) The micro-architecture of mitochondria at active zones: electron tomography reveals novel anchoring scaffolds and cristae structured for high-rate metabolism. J Neurosci 30:1015–1026
103. Premsler T, Zahedi RP, Lewandrowski U et al (2009) Recent advances in yeast organelle and membrane proteomics. Proteomics 9:4731–4743
104. Rasmusson AG, Geisler DA, Moller IM (2008) The multiplicity of dehydrogenases in the electron transport chain of plant mitochondria. Mitochondrion 8:47–60
105. Rebeille F, Alban C, Bourguignon J et al (2007) The role of plant mitochondria in the biosynthesis of coenzymes. Photosynth Res 92:149–162
106. Reddy PH (2009) Amyloid beta, mitochondrial structural and functional dynamics in Alzheimer's disease. Exp Neurol 218:286–292
107. Rittscher J (2010) Characterization of biological processes through automated image analysis. Annu Rev Biomed Eng 12:315–344
108. Rizzuto R, Pinton P, Carrington W et al (1998) Close contacts with the endoplasmic reticulum as determinants of mitochondrial Ca^{2+} responses. Science 280:1763–1766
109. Rossignol R, Gilkerson R, Aggeler R et al (2004) Energy substrate modulates mitochondrial structure and oxidative capacity in cancer cells. Cancer Res 64:985–993
110. Rowland KC, Irby NK, Spirou GA (2000) Specialized synapse-associated structures within the calyx of Held. J Neurosci 20:9135–9144
111. Rust M, Bates M, Zhuang X (2006) Sub-diffraction-limit imaging by stochastic optical reconstruction microscopy (STORM). Nat Methods 3:793–795
112. Saraste M (1999) Oxidative phosphorylation at the fin de siecle. Science 283:1488–1493
113. Scaduto RC Jr, Grotyohann LW (1999) Measurement of mitochondrial membrane potential using fluorescent rhodamine derivatives. Biophys J 76:469–477
114. Schauss AC, Bewersdorf J, Jakobs S (2006) Fis1p and Caf4p, but not Mdv1p, determine the polar localization of Dnm1p clusters on the mitochondrial surface. J Cell Sci 119:3098–3106
115. Scheffler IE (2001) Mitochondria make a come back. Adv Drug Deliv Rev 49:3–26
116. Scheffler IE (2008) Mitochondria. Wieley, New Jersey
117. Schmidt R, Wurm CA, Jakobs S et al (2008) Spherical nanosized focal spot unravels the interior of cells. Nat Methods 5:539–544
118. Schmidt R, Wurm CA, Punge A et al (2009) Mitochondrial cristae revealed with focused light. Nano Lett 9:2508–2510

119. Schubert W, Bonnekoh B, Pommer AJ et al (2006) Analyzing proteome topology and function by automated multidimensional fluorescence microscopy. Nat Biotechnol 24:1270–1278
120. Scorrano L, Ashiya M, Buttle K et al (2002) A distinct pathway remodels mitochondrial cristae and mobilizes cytochrome *c* during apoptosis. Dev Cell 2:55–67
121. Shariff A, Kangas J, Coelho LP et al (2010) Automated image analysis for high-content screening and analysis. J Biomol Screen. doi:10.1177/1087057110370894
122. Sheahan MB, McCurdy DW, Rose RJ (2005) Mitochondria as a connected population: ensuring continuity of the mitochondrial genome during plant cell dedifferentiation through massive mitochondrial fusion. Plant J 44:744–755
123. Sheridan C, Delivani P, Cullen SP et al (2008) Bax- or Bak-induced mitochondrial fission can be uncoupled from cytochrome *c* release. Mol Cell 31:570–585
124. Smiley ST, Reers M, Mottola-Hartshorn C et al (1991) Intracellular heterogeneity in mitochondrial membrane potentials revealed by a J-aggregate-forming lipophilic cation JC-1. Proc Natl Acad Sci USA 88:3671–3675
125. Stevens B (1977) Variation in number and volume of the mitochondria in yeast according to growth conditions. A study based on serial sectioning and computer graphics reconstitution. Biol Cell 28:37–56
126. Sträuber H, Müller S (2010) Viability states of bacteria-specific mechanisms of selected probes. Cytometry A. doi:10.1002/cyto.a.20920
127. Szabadkai G, Simoni AM, Bianchi K et al (2006) Mitochondrial dynamics and Ca^{2+} signaling. Biochim Biophys Acta 1763:442–449
128. Thomas B, Beal MF (2007) Parkinson's disease. Hum Mol Genet 16(2):R183–194
129. Wang C, Youle RJ (2009) The role of mitochondria in apoptosis. Annu Rev Genet 43:95–118
130. Warren G, Wickner W (1996) Organelle inheritance. Cell 84:395–400
131. Weber K, Ridderskamp D, Alfert M et al (2002) Cultivation in glucose-deprived medium stimulates mitochondrial biogenesis and oxidative metabolism in HepG2 hepatoma cells. Biol Chem 383:283–290
132. Wikstrom JD, Katzman SM, Mohamed H et al (2007) beta-Cell mitochondria exhibit membrane potential heterogeneity that can be altered by stimulatory or toxic fuel levels. Diabetes 56:2569–2578
133. Wikstrom JD, Twig G, Shirihai OS (2009) What can mitochondrial heterogeneity tell us about mitochondrial dynamics and autophagy? Int J Biochem Cell Biol 41:1914–1927
134. Wurm CA, Jakobs S (2006) Differential protein distributions define two subcompartments of the mitochondrial inner membrane in yeast. FEBS Lett 580:5628–5634
135. Yamaguchi R, Lartigue L, Perkins G et al (2008) Opa1-mediated cristae opening is Bax/Bak and BH3 dependent, required for apoptosis, and independent of Bak oligomerization. Mol Cell 31:557–569
136. Yamaguchi R, Perkins G (2009) Dynamics of mitochondrial structure during apoptosis and the enigma of Opa1. Biochim Biophys Acta 1787:963–972
137. Youle RJ, Karbowski M (2005) Mitochondrial fission in apoptosis. Nat Rev Mol Cell Biol 6:657–663
138. Zellnig G, Zechmann B, Perktold A (2004) Morphological and quantitative data of plastids and mitochondria within drought-stressed spinach leaves. Protoplasma 223:221–227

Adv Biochem Engin/Biotechnol (2011) 124: 21–54
DOI: 10.1007/10_2010_83

Published Online: 17 November 2010

Advanced Microscopy of Microbial Cells

Janus A. J. Haagensen, Birgitte Regenberg and Claus Sternberg

Abstract Growing awareness of heterogeneity in cells of microbial populations has emphasized the importance of advanced microscopy for visualization and understanding of the molecular mechanisms underlying cell-to-cell variation. In this review, we highlight some of the recent advances in confocal microscopy, super-resolution optical microscopy (STED, SIM, PALM) as well as atomic force microscopy and Raman spectroscopy. Using examples of bistability in microbial populations as well as biofilm development and differentiation in bacterial and yeast consortia, we demonstrate the importance of microscopy for visualization of variation between cells in phenotypic traits such as gene expression.

Keywords Advanced microscopy techniques · Single-cell gene expression · Bistability · Biofilm development and differentiation · Bacteria · Yeast

Abbreviations

AFM	Atomic force microscope
AOBS	Acousto-optical beam splitter
AOTF	Acousto-optical tunable filter
CARS	Coherent anti-Stokes Raman spectroscopy
CFP	Cyan fluorescent protein
CLSM	Confocal laser scanning microscope
EPS	Extracellular polymeric substance
FRAP	Fluorescence recovery after photobleaching

J. A. J. Haagensen
Stanford University, The Bio-X Program, Clark Center,
318 Campus Drive, Stanford, CA 94305, USA
e-mail: Haagensen@stanford.edu

B. Regenberg and C. Sternberg (✉)
Department of Systems Biology, Technical University of Denmark,
Building 301, 2800 Kongens, Lyngby, Denmark
e-mail: cst@bio.dtu.dk

B. Regenberg
Department of Biology, Molecular Integrative Physiology,
Universitetsparken 13, 2100 Copenhagen, Denmark
e-mail: bregenberg@bio.ku.dk

GFP	Green fluorescent protein
MP	Multi photon
NA	Numerical aperture
PALM	Photo-activated localization microscopy
PE	Polyethylene
PI	Propidium iodide
PMT	Photo multiplier
PP	Polypropylene
PVC	Polyvinyl chloride
SIM	Structured illumination microscopy
SERS	Surface enhanced Raman spectroscopy
STED	Stimulated emission depletion
TERS	Tip enhanced Raman spectroscopy
STORM	Stochastic optical reconstruction microscopy
YFP	Yellow fluorescent protein

Contents

1 Introduction

While culture or community-based approaches where millions of organisms are studied collectively as a single entity are adequate for general physiological studies, microscopy of single cells and subpopulations has lately been realized as an important and sometimes indispensable tool when a community of apparently homogenous cells is investigated for phenotypic diversity.

The often used assumption when monitoring gene expression in a bacterial population is to consider the population as a uniform pool. However, this will only report average values and never pick up variations between individual single cells. This assumption has been prevailing as analyzing and visualizing single cells in large populations have been difficult or impossible.

Microscopy has been an important method for analysis of microbial cells since Antonie van Leeuwenhoek described amoebae and other microorganisms in the late 1600s. The ‘compound microscope’, i.e., a microscope with more than one lens, was invented almost a century before these first reports of living microorganisms [8]. Development of the microscope has since undergone numerous improvements, where some of the most notable are the invention of the fluorescence microscope [60], the phase contrast principle [150], the electron microscope in 1931 [114], the confocal microscope in 1955 [89], and the atomic force microscope (AFM) in 1986 [15] and derivatives of these.

Today, new technologies provide the possibility to differentiate between subpopulations and to analyze single cells for altered gene expression using advanced microscopy, cell sorting and a growing number of fluorescent reporter tools. An increased awareness of cellular differentiation has emphasized the need for understanding of microbial gene expression and provides a more efficient and direct treatment of bacterial and fungal infections [128, 138]. Furthermore, a better insight into the regulation of differentiation in bacteria and yeasts can lead to a wider application in biotechnology [27].

Microbes are found everywhere from settings in the human host to soil and aquatic environments. They are constantly meeting new environmental conditions and have evolved highly sophisticated abilities to adapt to changes. Several adaptive mechanisms are used by microbial populations to turn on and off genes stochastically in a population of cells while being in the same environment. This approach can ensure that a subpopulation of cells will survive in a situation where life conditions suddenly change (bet-hedging). Bistability is an example of a molecular mechanism that has evolved to diversify the transcriptional program and phenotype of a clonal population within the same environment. Bistability can be described as an inheritable and reversible switch at the level of transcription that does not involve genetic rearrangements or mutations but is rather epigenetic in its nature [27, 35, 39].

In this chapter, we will describe the current state of the art microscopy techniques and use bistability in microbial populations and biofilm development and differentiation in these as examples of biological properties, which to a large extent has been observed and investigated using advanced microscopy.

2 Advanced Microscopy and Tools

2.1 The Standard Confocal Microscope Today

From the time the first confocal microscopes became commercially available, a long range of improvements has been implemented and are now standard: The lasers used in most microscopes have a much improved lifetime compared to early models, and in particular the ruggedness of diode lasers significantly extend the period before replacement is necessary. In addition, this type of laser requires less for the installation environment in terms of power supply and cooling—all factors that lower the running costs. Traditional gas lasers, such as the most common Argon gas laser, has also improved and now provides longer life spans. The detectors of confocal microscopes are traditionally photo-multipliers, PMTs. Their sensitivity largely determines the overall image quality, particularly in conditions of low light intensity. Newer PMTs have improved signal-to-noise ratios, enabling detection of fainter fluorescence signals. The band-pass filters on the emission side has also been improved from the original glass filters, which were placed in the light path manually, over motorized filter wheels to the acousto-coupled beam deflector [42], commonly known as an acousto-optical tunable filter (AOTF). AOTF selectively deflects specific wavelengths out of the light path, leading only the desired emission to the detectors. This technology has been developed into multiline spectral splitters, such as the acousto-optical beam splitter (AOBS) [16], which allows for detection of multiple wavelength ranges simultaneously. An alternative to this is the technology where the light beam is spectrally split by a grate working like a prism, and selected parts of the spectrum are captured by an array of detectors.

Confocal microscopes have been and are widely used in the study of complex microbial communities (biofilms) since the first reports by the Caldwell group in 1992 [21, 85] and onto the present day [97, 103, 148].

2.2 Wide Spectrum ("white") Lasers

Traditional gas lasers can emit light of at most a few, well-defined wavelengths. For instance, the most commonly used laser in confocal microscopy, the Argon gas laser, emits a range of monochromatic wavelengths at 488, 351, 454.6, 457.9, 465.8, 476.5, 496.5, 501.7, 514.5 and 528.7 nm. However, in the standard visible light configuration, the laser is configured to mainly provide 488-nm blue excitation light. Some of the other wavelengths require alteration of the mirror system to be usable, and some are considerably weaker than the 488-nm line. It is also common that commercial instruments have multiple lasers with different characteristics to enable the use of more fluorophores. Despite multiple simultaneous lasers, the choices of excitation wavelengths are limited. To alleviate this, the super-continuum laser, also known as the "white" laser, has been introduced. The super-continuum laser is a compound laser consisting of a pump laser and a crystal photonic fiber optic. A conventional laser delivers a narrow band laser illumination to the end of the fiber that consists of a bundle of hollow tubes. When passing through the fiber the spectrum is broadened, resulting in a wide spectrum emission. The width of the spectrum depends on the pattern and length of the tube arrangement inside the fiber. The first verified reports of broad spectrum laser light were published already in 1970 [3, 4]. However, it was only after the invention of the hexagonal photonic fiber that the white laser we know today was made useful for practical applications [81]. The technology is currently used for confocal microscopy [18] and stimulated emission depletion (STED) [145–147] (see below). A typical wide spectrum laser used in confocal microscopy has a spectrum covering most of the visible wavelengths, 470–670 nm.

2.3 Multi-Photon/Two-Photon Microscopy

Conventional confocal imaging has its basis in the fluorescence microscope, i.e., the specimen must contain a dye that is fluorescent—an added dye in the form of a chemical, a fluorescent protein expressed by cells in the sample or auto-fluorescence. The sample is illuminated using a laser throughout the entire depth, and fluorescence is emitted from the whole cross section. The emission pinhole will remove most of the fluorescence from all planes except the focal plane. This method is most likely harmful to living cells as all cells are exposed to laser radiation for the duration of the scanning, and the aberrant fluorescence from the layers other than the focal plane *do* in fact contribute to noise in the image.

The two- or multi-photon (MP) principle [51] predicts that when two or more photons hits a fluorophore simultaneously (i.e., within a femtosecond timeframe) the two photons will both contribute to the excitation. In other words, a longer wavelength laser illumination can provide a localized energy pulse corresponding to that of a shorter, more energy-rich wavelength used in a conventional one-photon system. Thus, using infrared laser light, it is possible to excite molecules that require, e.g., blue light for excitation. Since the multi-photon effect only occurs where more photons are precisely in synchronicity, it is possible to exploit this property to narrow the illumination to an extremely small volume. This is used in the two-photon confocal microscope [36]. The MP-confocal microscope gives better vertical (z-)resolution, about 100 nm [96] compared to 5–700 nm for conventional, one-photon confocal microscopy. High-resolution confocal microscopy of relatively thick specimens is possible with MP excitation [95, 96]. The emission pinhole is not necessary, since only objects in a small volume in the focal plane are

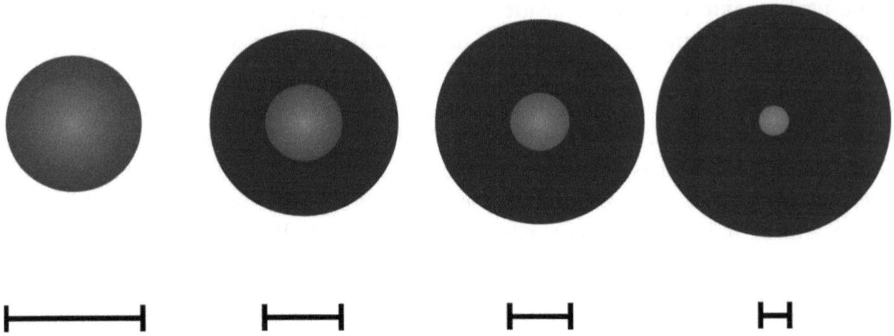

Fig. 1 Stimulated emission depletion (STED) microscopy. Without STED the fluorescence emission light (*light grey*) produces a light spot, which cannot be localized to a better resolution than predicted by Abbe. By application of a ring-shaped STED laser light (*dark grey*) that extinguishes parts of the fluorescence from the sample, the effective resolution is improved. By increasing the STED laser effect, the resolution improves accordingly, as indicated by the *size indicator bars*

excited by more than one photon at a time, the remainder of the specimen only experiences long wavelength light with less detrimental effects (both for cytotoxicity and bleaching). However, since the light flux is very high in the small volume that is excited, cell damage can occur if care is not taken to protect the exposed cells [67]. MP microscopy in life sciences was reviewed by König [83].

2.4 Super Resolution Confocal Microscopy

Ernst Abbe (1840–1905) determined the theoretical optical resolution as $d = \lambda/2\mathrm{NA}$, where λ is the wavelength and NA is the numerical aperture of the objective [1]. Using visible light, this means that the practical horizontal resolution (the distance to resolve two objects) is 200–250 nm. While this is sufficient for many purposes, analysis of sub-cellular structures, surface components and appendages is rarely possible. Electron microscopy and atomic force microscopy (see below) have several-fold improved resolution but have other shortcomings; e.g., electron microscopy usually requires the sample to be fixed, dried and dyed with metal dyes, whereas atomic force microscopy is restricted to analysis of the top surface of structures with little variation in height. The optical microscope is superior since it allows scrutiny of living, wet samples in three dimensions.

A number of methods now exist to circumvent the law of Abbe, either as a physical, on-line method or as a computational reconstruction from several interlaced images. Probably the most prominent high-resolution technology is STED [40, 57, 63, 79]. In STED the sample is illuminated by two tightly synchronized light pulses. The fluorophore is excited with a pulsed excitation beam, e.g., 640 nm, which causes the sample to emit light (Fig. 1). Without STED, the emitted light will have a diffraction limit that in part is determined by Abbe's equation. STED adds an additional light pulse at, e.g., 730–780 nm, which is doughnut shaped with a dark center. The wavelength of the STED illumination does not excite the fluorophore—rather it causes the already excited molecules to return to the ground state without emitting light. The result is that fluorescence is only emitted from the dark center, the size of which is determined by the laser power of the STED light, increasing the effective resolution to this center area.

Instrument development platforms have reached a resolution of 5.8 nm [111], while commercial instruments typically will have a lower effective resolution. The first generation STED microscopes were limited in the number of fluorescent dyes that could be employed since the laser configuration required was limited to 640 nm excitation and 730–780 nm depletion. Using lasers with tunable excitation wavelengths, the STED technology has broadened the versatility of the instrument. Using the continuous wave laser or super-continuum lasers, it is possible to utilize commonly used markers such as green fluorescent protein (GFP) with a resolution of 29–60 nm [61, 145–147].

2.5 Other Microscopy Techniques Surpassing the Diffraction Resolution Limit

Computational treatment of images after acquisition demonstrate other methods for rendering of images with sub-diffraction resolution. Three such methods have been commercialized, the stochastic optical reconstruction microscopy (STORM), photo-activated localization microscopy (PALM) and structured illumination microscopy (SIM).

STORM is based on sequential excitation of photo-switchable fluorophores in a sample followed by reconstruction of a high-resolution image from a series (sometimes hundreds or thousands) of images of the same field of view [68, 115]. The principle of STORM is that only a fraction of the fluorophores are excited at any time in the sample, which then is recorded by the microscope. A quenching light pulse extinguishes the fluorophores and excites another set of dye spots in the sample. By changing the position that is illuminated or quenched, it is possible to record a fine map of the position of the fluorescent molecules. Using computer programs, this information can be converted into an image with in principle unlimited resolution [115], but mechanical limitations give an effective resolution in the range of 20 nm. PALM is an independently developed technique using the same principle as STORM [13, 64, 104, 122]. One prominent feature of this method for resolution improvement is that that in principle standard microscopy equipment and a low cost laser are all that is required of the hardware, although advanced software is needed for the post-recording image manipulation. STORM and PALM have been used for 3D imaging [68, 76, 105, 123, 137] and multicolor imaging [10, 118, 122].

SIM uses the principle of interlaced information from structured illumination by very high-resolution patterns that are projected onto the fluorescent sample. When two patterns are overlaid, these images result in a moiré fringe with a lower resolution. Figure 2 demonstrates how two idealized high-resolution patterns result in a moiré fringe (here perceived as horizontal lines) with lower resolution (spacing) than the original patterns. In the SIM one of the patterns is comprised of the structure in the sample (the distribution of fluorescent dye) and the overlaying pattern of the illumination pattern. Using multiple images recorded with different illumination angles and mathematical processing, it is possible to reconstruct an image with resolution corresponding to the resolution of the illumination pattern.

This method gives approximately a two-fold improvement in resolution compared to the diffraction limit determined by Abbe [48]. A recent extension of this technique, saturated SIM (SSIM), has further moved the resolution limit. SSIM utilizes nonlinear patterned excitation of fluorescent samples and the image reconstruction techniques of SIM to achieve in principle unlimited resolution, while practical experimental setups have yielded a resolution of roughly 50 nm, or four times that of standard microscopy [49, 62]. Due to the requirement for overlay of several sequentially recorded images, SIM and SSIM are less well suited for imaging of living cells, but a recent report has demonstrated SIM of slowly moving cells [66].

2.6 Atomic Force Microscopy

Atomic force microscopy is a non-optical surface scanning method in which a tip (probe) at the end of a flexible cantilever is transversing a structured surface. The AFM can operate in several different modes. The two most commonly used for imaging are contact mode and tapping mode, or dynamic mode. In contact mode, the tip is dragged across the surface and irregularities on the surface cause the cantilever, which is carrying the tip, to bend up and down. A laser is recording the bend, which can be directly correlated to the topology of the surface. In tapping mode an oscillating frequency is applied to the cantilever, making the tip move up

Fig. 2 Structured illumination microscopy (SIM). In consecutive scans, a patterned illumination is applied to the sample. The illumination pattern in combination with the fluorescence pattern in the sample (*symbolized by the lines in the middle panel*) result in a moiré pattern that can be the basis for computer-assisted image reconstruction

and down towards the surface. In tapping mode the tip is subjected to a combination of attractive and repulsive forces, which influence the amplitude of the oscillation. A feedback loop corrects the distance between the cantilever and the sample by piezo-electrical actuators (moving the stage or the cantilever-mount) to bring the amplitude back to the initial state (Fig. 3). The required correction can be correlated to the distance between the tip and sample at a specific position and then converted to a height (3D) image of the surface when the tip is moved across the sample.

The AFM has atomic resolution on crystalline surfaces and nanometer resolution on other surfaces. It can be operated in liquids and gaseous environments. Applications of atomic force microscopy in biological research began in the early 1990s [50, 86, 117, 133]. As the AFM became more integrated in life science research, this new tool provided a new approach for the examination of biomolecules including proteins [55, 107, 117], DNA [56, 86, 134] and highly topographic samples, such as bacterial [5, 17, 19, 22, 112] and yeast cells [44, 77] at nanoscale resolution. Most importantly, samples could be imaged in physiological relevant media, and in the case of bacteria and mammalian cells living cells could be imaged in their native environment. While the AFM provides imaging with extreme resolution, it does only facilitate analysis of surfaces that are accessible from above; hence, it is not suitable for analysis of intercellular processes.

2.7 Single-Cell Fluorescent Labeling, Visualization and Physiology

Several microscopic methods require that cells are fluorescent. Consequently, it is important to have tools available for staining or marking investigated cells with specific fluorescent labels.

Recently, there has been a development of fluorescent stains, such as the Syto stains (Invitrogen, Carlsbad, CA), that efficiently, although unspecifically, can

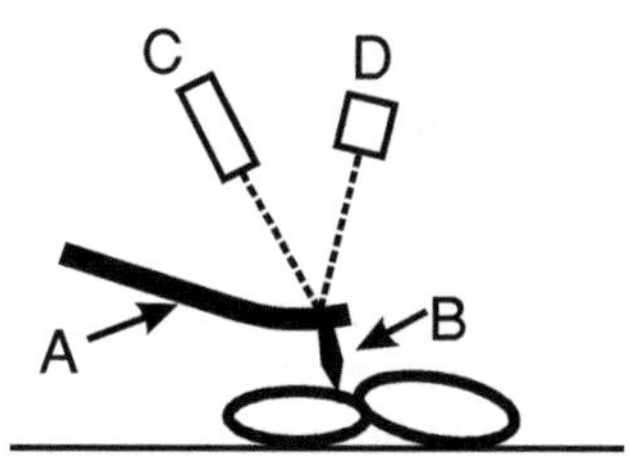

Fig. 3 Atomic force microscopy. The sample is scanned by a moving tip (**b**), which is attached to a flexible cantilever (**a**). The deflection of the cantilever is recorded by using a laser (**c**) illuminating a spot on the back of the cantilever. The position of the spot is recorded by a photomultiplier (**d**), and a feedback loop moves the cantilever or the sample to return the deflection to a neutral state (as determined by the location of the laser spot)

stain cells. Combinations of stains with different excitation and emission wavelengths are available for possible use together with reporter gene constructs. Using Syto9-labeled cells in combination with propidium iodide (PI), it is possible to specifically determine living and dead cells in a population. The dye Syto9 will mark all cells green, while it is generally assumed that only cells with a damaged membrane integrity will be stained by the red PI dye, indicating dead cells. PI will reduce Syto9 in dead cells, making them only fluoresce red. Recent results suggest that propidium iodide might be of limited use as a cell viability indicator in some settings and for some strains. Therefore, it is important for each species and environment to calibrate the concentration of dye [91, 121].

As an alternative to direct staining, a common method used today is to modify the cells of interest genetically by chromosomal tagging with a gene cassette encoding a fluorescent protein or by plasmid introduction. In this way GFP (green), RFP (red), CFP (cyan) and YFP (yellow) have been successfully introduced into many different cell types. GFP-tagged cells can substitute the use of Syto9 in the live/dead assay described above.

Fluorescent tagging can be used as simple labeling to verify and visualize the location of several species in a mixed community. By selecting suitable variants of fluorescent protein genes and promoters, this kind of tagging can be used for monitoring gene expression in specific cells. This way metabolic/physiological activity has been determined in biofilms by introducing constructs encoding for GFP derivatives with a short half-life, placed under transcriptional control of a ribosomal promoter. Cells that have a high activity will show as bright green, whereas cells with low or no activity show little or no fluorescence [130].

2.8 Raman Microscopy

Raman spectroscopy is a method that can produce a fingerprint of the chemical composition of materials in a cell based on Raman scattering of the molecules in the materials. Molecules that are hit by an incoming photon can either absorb or scatter the light, or not interact at all with the light. The scattered light will primarily have the same wavelength as the incoming light, whereas a very small fraction (1 per 10^6–10^8 photons) will have a different, higher or lower, wavelength due to vibrational or rotational effects in the molecule: When a photon interacts with the electron cloud and bonds in a molecule, it can excite it to a more energetic state. Most photons excite a molecule to a higher virtual energy state from a relaxed state. When the molecule returns to the relaxed state, energy of the same magnitude (and hence photons with the same wavelength) as the excitation energy is released. This is called Rayleigh scattering (Fig. 4).

Occasionally, however, the molecule may be excited to the higher virtual energy state and return to an energy state that has a higher level than the relaxed state, releasing less energy than the excitation photon. This will then result in a scattering photon with less energy, i.e., longer wavelength, called Stokes

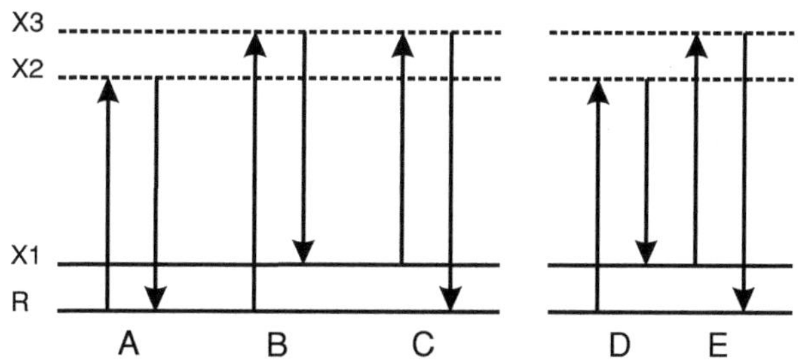

Fig. 4 Energy levels. Incident photons excite molecules to higher virtual energy states. **a** Rayleigh scattering. **b** Stokes scattering. **c** Anti-Stokes scattering. **d**, **e** Coherent anti-Stokes Raman radiation. **d** Fixed wavelength pump laser excites to X2; the Stokes laser beam facilitates relaxation to X1. **e** The probe laser beam excites molecules at X1–X3, with subsequent relaxation to the ground state. *X1* Vibrational energy state. *X2*, *X3* Virtual energy states. *R* relaxed or ground state

scattering. Similarly, a condition can occur where the molecule is already in a higher virtual energy state when the incoming photon excites the molecule. If the molecule subsequently returns to the relaxed state, the scattered photon will have a higher energy than the incoming photon, resulting in a shorter wavelength. This very rare reaction is called anti-Stokes scattering. Together Stokes and anti-Stokes scattering is called Raman scattering. Raman spectroscopy utilizes a single frequency of radiation for excitation, and the spectrum of frequency shifted emission from the sample is the Raman emission spectrum. The Rayleigh scattering is filtered to leave only the much weaker Raman scattering [126]. Every molecule will result in a characteristic Raman spectrum, which is the result of the combined Raman scattering of the molecular bonds and electron clouds in that particular compound. A complex organism will hence give a complicated spectrum consisting of the overlaid combined spectra from all the molecules (e.g., proteins, fatty acids, nucleic acids, etc.) in the sample, where signature peak heights and positions are representative for individual components of the cells [94]. This can be used to follow the change in chemical composition of single individual cells as a result of, e.g., growth rate or interaction with the environment, and in this way differentiation in a population can be determined on a single-cell level (see examples later in this chapter). Raman spectroscopy can be combined with confocal microscopy to provide 3D information on cell identity or chemical composition of, e.g., extracellular substances (EPS) in biofilms [106, 116, 144].

The Raman scattering is inherently very weak, and a number of methods exist to enhance sensitivity. Surface-enhanced Raman spectroscopy (SERS) is frequently used to amplify the weak Raman signals. In SERS the sample is placed on a typically silver- or gold-covered surface. The physical explanation for the enhancement (up to 10^{11} fold) is not fully elucidated, but it is believed that an increase in the electrical field due to the excitation of the gold or silver surface plasmons by the laser light source boosts the Raman scattering intensity [75]. SERS has successfully been used for identification and characterization of bacteria

[73, 74]. In combination with the AFM, utilizing metal-coated tips, the Raman measuring capability can be combined with very high resolution, enabling chemical mapping of surfaces, e.g., of bacteria down to a molecular scale. In this technique, termed tip-enhanced Raman spectroscopy (TERS), the Raman excitation is performed at the metal-coated AFM tip, linking the atomic force microscopy high-resolution imaging to very localized SERS [20, 98, 100]. Further signal enhancement can be achieved using coherent anti-Stokes Raman spectroscopy (CARS). This technique relies on two laser excitation sources. These two lasers, the fixed frequency pump laser and the tunable Stokes laser, simultaneously excite the molecule to virtual state (X2) and vibrational state (X1) (see Fig. 4d, e): When the Stokes laser has the right frequency, the return from X2 to a lower vibrational energy state (X1) occurs via a stimulated Stokes emission. When the sample molecule is in this state, it can be further excited by a probe laser beam to the higher virtual energy state (X3) (in the actual setup this beam is provided by the Stokes laser). When the molecule relaxes to the ground state (R), it emits a photon with a higher energy than the excitation photon, resulting in an anti-Stokes effect, i.e., a higher frequency [11, 90]. This setup produces coherent anti-Stokes photons resulting in a dramatically enhanced signal. Furthermore, since the emitted Raman scattering has a shorter wavelength than the excitation photons, interference from fluorescence is eliminated [90]. A complication is that so-called phase matching is required to conserve the sum of photon momentum; it means that the strong confined beam of CARS photons is being emitted in its own direction, depending on the direction of the two incoming beams and their frequencies [23]. This method has already been used for imaging of bacterial cells, although at an experimental stage [25, 153]. The emergence of new commercial instruments with CARS-enhanced confocal microscopy gives great promise for the future coupling of structure and chemical composition of microorganisms.

2.9 Presentation and Analysis Software

Many software packages exist for the analysis of microscopy images. They can roughly be divided into: general purpose and dedicated programs, programs for presentation of single images or image stacks, and programs for analytical purposes. Representative examples are provided in Table 1.

3 Bacterial Single-Cell and Biofilm Microscopy, Bistability and Subpopulations

The following sections will concentrate on the phenomenon of bistability in bacterial populations and how microscopy of single cells and biofilm structures has revealed differentiation between cells in a population. *Bacillus subtilis* and

Table 1 Examples of software packages for microscope image processing. Image processing programs

Name	General purpose	Dedicated (instrument specific)	Analytical	2D	3D	Source
ImagePro Plus	X	X[a]	X	X	X	http://www.mediacy.com
ImageJ and NIH Image[b, c]	X		X	X	X	[2]
Zeiss ZEN (free viewer available)		X	X[a]	X	X	Carl Zeiss AG, Germany
Leica LAS (free viewer available)		X	X[a]	X	X	Leica Microsystems, Germany
ISA[d]			X	X	X	[87, 149]
ISA3D[d]			X	X	X	[14, 87]
COMSTAT and COMSTAT2[b]			X		X	[65] http://www.comstat.dk
Daime[b]	X		X	X	X	[34]
Imaris	X	X[a]		(X)	X	http://www.bitplane.ch
Phlip[b]			X	X	X	http://www.phlip.org
MicrAn and ConAn	X		X	X	X	http://www.biocon-online.de

[a] Via add-on module
[b] Free (public domain)
[c] Open source
[d] Algorithm source code available

Escherichia coli will be used as the main case studies of bistability, whereas *Pseudomonas aeruginosa* will be used as a model organism in connection to differentiating subpopulations found in biofilms and visualized using confocal microscopy.

3.1 Bistability in B. subtilis

3.1.1 Competence

By recent advances in the ability to investigate gene expression on a single-cell level, several studies have demonstrated bistability and differentiation in gene expression. Under certain growth conditions *B. subtilis* cells are able to enter the state of genetic competence, i.e., the bacteria can take up free DNA and incorporate it into their genome by recombination via the process of transformation [38]. When *B. subtilis* enters the stationary phase, competence occurs naturally in a subpopulation of cells and depends this way on the growth phase of the single cell as well as nutrient composition and availability in the near environment [37].

Competence development in *B. subtilis* is regulated by the transcriptional activator ComK, which binds to the *comK* promoter and thus acts as an auto regulator [136]. Fluorescent microscopy of single cells with GFP expression controlled by the *comK* promoter showed that the population divides into two subpopulations when entering the stationary phase, this way demonstrating bi-stability in competence expression; some cells express *comK* while others do not activate the ComK promoter [92, 127].

When cells are growing exponentially the level of ComK is low as the housekeeping protease complex MecA-ClpC-ClpP is degrading the ComK protein. Furthermore, at least two repressors, Rok and CodY, act on the *comK* promoter. At near stationary phase, quorum sensing results in upregulation of ComS, which will partly block the inactivation of ComK by binding to the MecA-ClpC-ClpP complex [52, 92, 93].

3.1.2 Sporulation

Differentiation in cell types can in some cases be distinguished by cell morphologies using light microscopy, but in many cases a clear picture of different cell types in a population can only be analyzed by use of transcriptional reporter gene fusions for each cell type in combination with fluorescent microscopy, which also will make quantification by flow cytometry of the cell populations possible. Using time-lapse microscopy, it is furthermore possible on a single-cell level to follow dynamic development in growth activity, structure and gene expression using fluorescently labeled appropriate reporter strains. When *B. subtilis* reaches the stationary phase, nutrients become limited, and an endospore will be produced in some, but not all cells. Many cells lyse at this point, liberating nutrients, and the release of the endospore from mother cells eventually takes place. The released nutrients support a second growth period and production of spores. The produced spores will maintain their dormant stage, which is highly resistant to environmental stress until a preferred environmental signal triggers germination.

Like in the case of ComK, the regulatory protein Spo0A has been found to exhibit bistable behavior during sporulation. Using a *gfp* fusion to a promoter under the control of Spo0A, it was shown by fluorescent microscopy that the population revealed a heterogeneous distribution of cells, with only a subpopulation of cells expressing Spo0A [47].

Spo0A is under the control of a positive feedback loop through activation of its own promoter as well as a double repressor system. Spo0A represses *abrB* and AbrB represses *sigH*, an activator of *spo0A*. A high expression of *spo0A* thus is necessary to reduce the AbrB-mediated repression of *sigH* resulting in further expression of *spo0A*. It was demonstrated using CFP and YFP in promoter fusions to AbrB and SpoIIA that expression of the *abrB* and *spoIIA* genes is distinct in individual cells during sporulation, resulting in a bimodal expression profile [140].

Single-cell tracking was used by Veening et al. [139] to investigate three cell forms in *B. subtilis*: spore-forming cells, lysing cells and actively growing cells

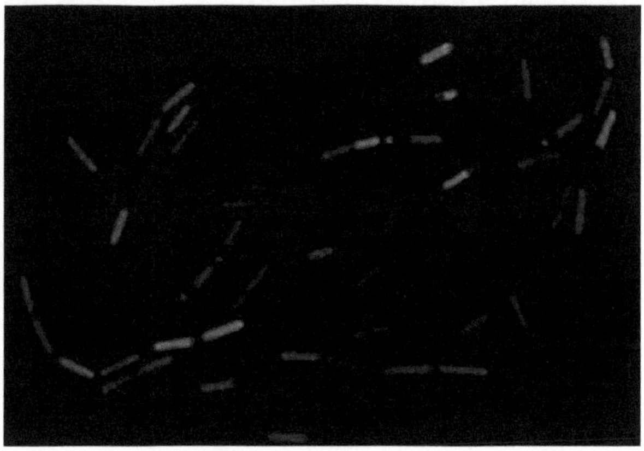

Fig. 5 Epigenetic inheritance in a *B. subtilis* sporulation, non-random bistability. The image shows how subfamilies of *spoIIA* Gfp expressing cells (*light grey*) develop a non-random bistability pattern [141]. Reproduced with permission. Copyright (2008) National Academy of Sciences USA

reappearing after cells had entered stationary phase. It was found that the fate of most cells already is determined before reaching the stationary phase. Cells that result in spore formation will not grow after the exponential phase, whereas cells that start growing actively after the first exponential phase will not become spore formers. A cell population that has lysis as its final cell fate will be able to join both groups of cells. By time-lapse recording of colony development in *B. subtilis*, using a GFP reporter system showing expression of SpoIIA, another regulatory protein regulating spore formation, it was visualized that spore-forming cells in most cases were situated next to each other (light grey cells in Fig. 5). This finding indicated a non-random development of the two subpopulations in a way that cell offspring's will share a phenotype with their parents in a bistable *B. subtilis* population [141].

3.1.3 Motility

Bistability in competence and sporulation in *B. subtilis* is pronounced during the late exponential and stationary growth phase, and fluorescent reporter systems and microscopy have been found useful for demonstrating this phenomenon. A third bistable regulation takes place in *B. subtilis*, namely in cell motility in exponentially growing cells. Only some cells express *sigD*, the sigma factor necessary for flagella production, and the result is a differentiating cell population of motile and non-motile cells. The non-motile cells will often appear as chain-like structures, making it possible to distinguish them from the motile cells even by morphology using light microscopy (Fig. 6) [78].

3.2 Bimodal Gene Expression and Biofilm Formation

For the last decade, it has been well know and accepted that bacteria in most environmental settings live in surface-associated communities called biofilms [28–

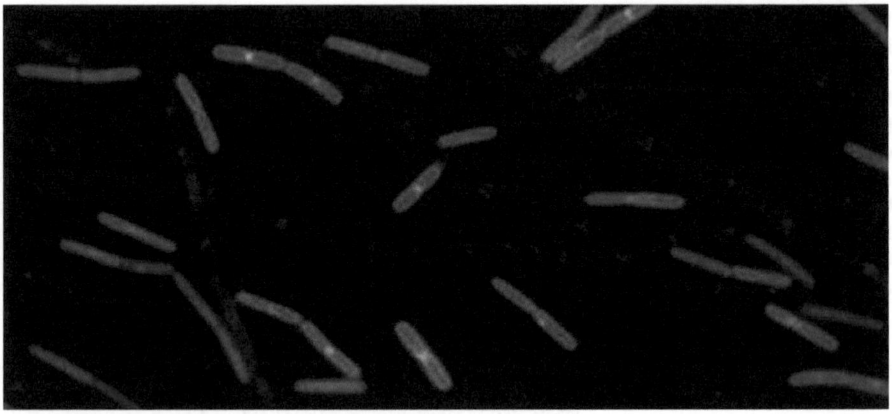

Fig. 6 Differentiation in motility. GFP was fused to the P(*hag*) promoter where *hag* is under control of the alternative sigma factor *sigD*, controlling motility and chaining in *B. subtilis*. Single cells were found GFP positive (*light grey*), whereas chains of cells did not show GFP expression [78]. Reproduced with permission from Cold Spring Harbor Press

32, 54, 82, 132, 135]. The importance of detailed knowledge about this bacterial lifestyle has proven to be required in order to understand many aspects of bacterial biology and is relevant both in environmental microbiology and medical microbiology. An important part of a biofilm structure in many cases is the presence of an extracellular polymeric substance (EPS) that is produced by the cells and that holds the biofilm structure together. EPS binding material between cells of *E. coli* was demonstrated using atomic force microscopy (Fig. 7) [29, 78].

Vlamakis and colleges [143] have combined gene expression profile studies with spatio-temporal differentiation in matrix-producing, motile and sporulating cells in connection to biofilm development and architecture of *B. subtilis*. In this study fluorescent protein reporter fusions were used to track expression of EPS matrix-producing, sporulating and motile cell types during dynamic biofilm development over a 72-h period. It was found after cell sorting of harvested biofilm cells that motility was upregulated in the initial stages of biofilm formation. That matrix-producing cells started to dominate after 24 h growth, and sporulating cells showed up in the older biofilm after 48 h. All three cell types coexisted in the mature *B. subtilis* biofilm, and production of EPS was carried out by only a subpopulation of cells [24].

Direct localization of the different cell types was also mapped during biofilm growth and maturation, and showed that motile cells dominated in the early biofilm at the top layers and represented only a minor part in the mature biofilm, mainly localized at the substratum. The matrix-producing cells were found randomly throughout the biofilm structure, whereas the sporulating cells were found in the upper structures in the mature biofilm. Using dual reporters it was shown that the mature biofilm harbored all three cell types and that motile cells were

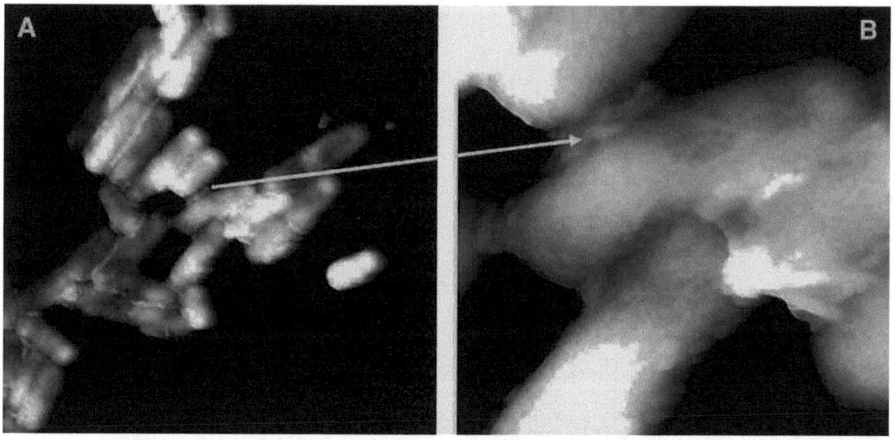

Fig. 7 AFM micrographs showing *E. coli* cells carrying the F plasmid, adhering to each other most likely by their F pili and by virtue of the produced EPS. **a** Cluster of cells and **b** zoomed image of a few cells in **a** showing the binding material between the cells (Haagensen and Sternberg, unpublished)

found mainly in distinct regions relative to sporulating cells that co-existed with the matrix-producing cells (Fig. 8).

Dual fluorescent-reporter systems combined with time-lapse microscopic recording was applied for the investigation of differentiation and transition of the three cell types at a single-cell level. By monitoring CFP-tagged cells reporting motility and YFP-tagged cells reporting matrix production, it was found that motile cells can perform a transition to matrix-producing cells. In the same way, it was shown that matrix-producing cells can turn into sporulating cells, whereas only a few motile cells have the ability to transform into sporulating cells (Fig. 9).

3.3 *Persistence and Bistability in* E. coli

Since the early days of antibiotic treatment of infections, it has been known that some bacteria develop resistance to certain antibiotics. The bacteria can cope with antibiotics by mutation, expressing efflux pumps, releasing indigenous antibiotic inactivation enzymes or the occasional acquisition of resistance-inferring genes from the environment. However, a subpopulation of cells in certain bacteria seems to have a different strategy in dealing with environmental stress. They develop persister cells with antibiotic tolerance, which is non-heritable and reversible, meaning that when the antibiotics are not present anymore the cells again become sensitive.

Until recently it has been difficult to isolate and investigate the small amount of persister cells developing in a population. Using microscopy in combination with microfluidic devices, phenotypic switching was studied in *E. coli*. The microfluidic

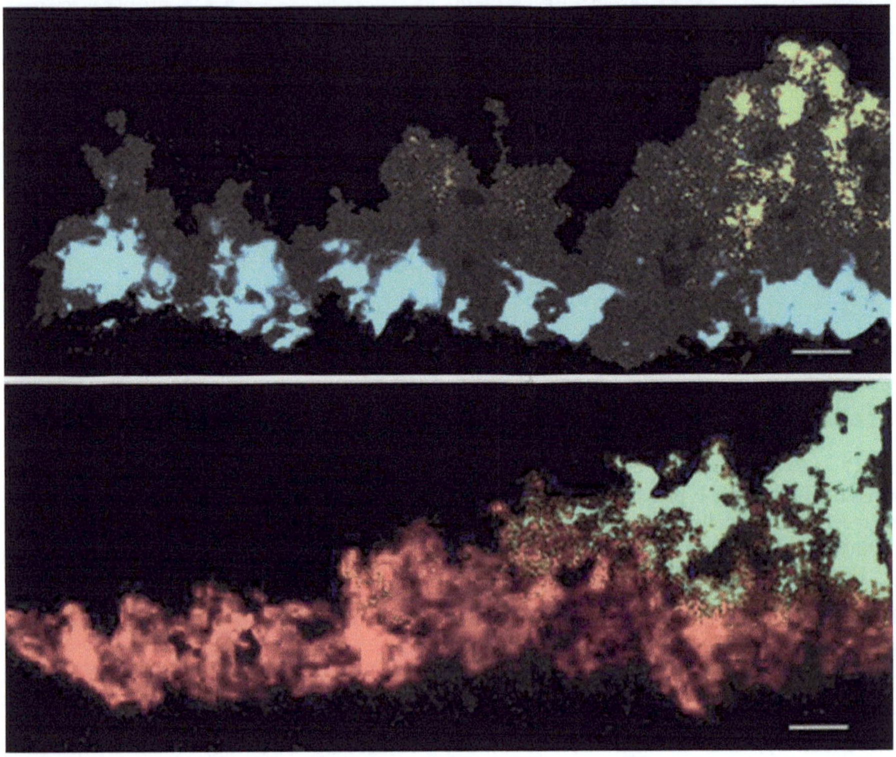

Fig. 8 Micrographs showing biofilm structure of *B. subtilis* and the distribution of motile cells in *blue*, matrix-producing cells in *red* and sporulating cells in *orange* [143]. Reproduced with permission from Cold Spring Harbor Press

channels were dimensioned to only allow propagation in one dimension (Fig. 10) [7]. The system thus allows for monitoring individual cells during growth and response to antibiotics, in this case ampicillin. It was possible to find persister cells already present before antibiotic challenge as a small subpopulation of cells showing a much reduced growth rate. During the antibiotic treatment only the persister cells survived in the microfluidic channels, and after removal of the antibiotic media the persister cells could resume growth (Fig. 10) [7].

Development of persister cells is a spontaneous bet-hedging survival strategy allowing *E. coli* to distribute its population heterogeneously such that some cells at all times are prepared for changing, adverse environmental conditions.

3.4 Raman Microscopy of Bacterial Cells

For *Clostridium* organisms, studies of the cell cycle have shown that cells germinating from spores develop rod-shaped cells, which eventually differentiate into

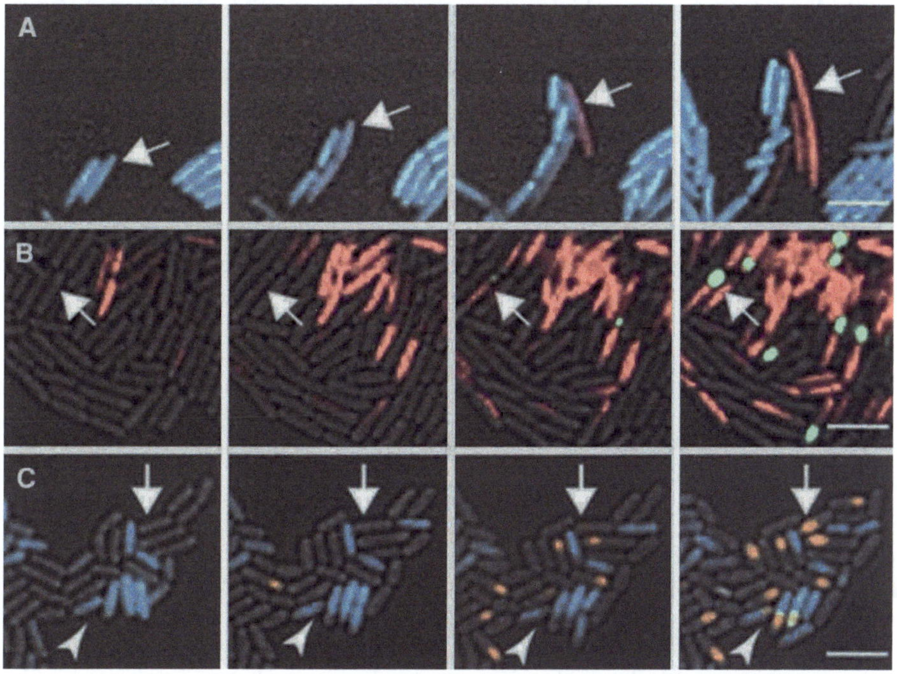

Fig. 9 Micrograph images showing dynamic development of single cells in a biofilm of **a** motile cells in blue and matrix-producing cells in *red*. The *arrow* shows a motile cell transitioning to a matrix-producing cell. **b** Matrix-producing cells in *red* and sporulating cells in *green*. The *arrow* shows matrix development followed by transitioning to a sporulation cell. **c** Motile cells in blue and sporulating cells in *orange*. The *arrow* shows that sporulating cells arise from non-motile cells (a few sporulating cells arise from motile cells, *arrowhead*) [143]. Reproduced with permission from Cold Spring Harbor Press

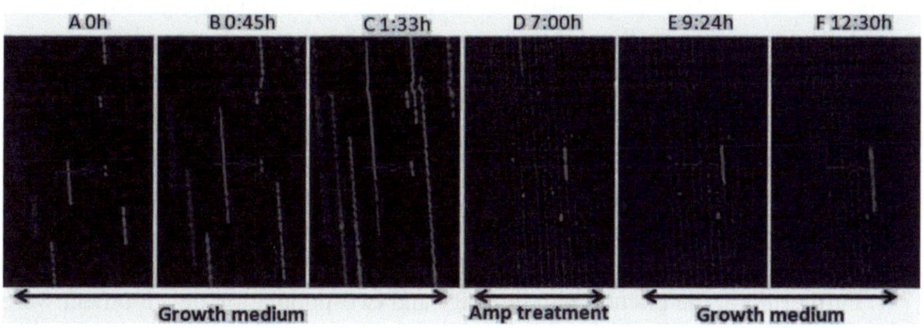

Fig. 10 Micrographs showing *E. coli* cells growing in a microfluidic chamber. Only slow-growing persister cells survived antibiotic treatment. **a–c** Cells are dividing and growing in narrow channels allowing the growth rate of the single cells to be monitored as the length of strings of cells developing over time. **d** Cells were exposed to ampicillin for 4 h. **e, f** After washing, cells were changed back to growth medium without ampicillin [7]. Reprinted with permission from AAAS

clostridial cell forms after which spores start to become visible. In this study it was demonstrated that Raman microscopy enables investigations of differentiation in cell physiology and composition on a single-cell level during the cell cycle of *C. acetobutylicum* [119]. In other examples, Raman microscopy has shown potential for investigation of the consequences of antibiotic treatment of bacterial cells [58, 99] and to differentiate between planktonic and biofilm-associated *Pseudomonas putida* cells [70]. Raman spectroscopy also has been used as a tool to determine identity of bacterial cells, although this requires a large training set (database) of spectra from already identified bacteria of the type to be determined, which by itself is not a trivial task [58, 113], reviewed recently by Harz [59].

3.5 **Pseudomonas aeruginosa** *Biofilm Development and Differentiating Subpopulations*

Pseudomonas aeruginosa has been used extensively for the study of microbial biofilm formation, and many of the most important contributions to our understanding of biofilm development comes from these studies. Lately, this organism has also been subject to differentiation studies and clinically relevant antibiotic treatment studies with the goal to investigate phenotypic heterogeneity in biofilm populations [9, 72, 80, 102, 148]. A combination of fluorescent reporter strains, confocal microscopy, fluorescent recovery after photo-bleaching (FRAP) and real-time microscopy has made it possible to follow the spatial distribution of phenotypic subpopulation development.

A mature biofilm of *P. aeruginosa* can be described as a mushroom-shaped structure composed of two major subpopulations, a subpopulation situated close to the substratum and a subpopulation forming the top of the mushroom [80]. The diversification into two subpopulations was shown by combining growth of GFP tagged *P. aeruginosa* in flow chambers with confocal time lapse microscopy in combination with FRAP. The GFP signal from cells in a section of a mature microcolony was bleached using high-intensity laser illumination in a rectangular region along the biofilm structure. The bleached area was monitored at small time intervals with respect to GFP fluorescent single-cell appearance. It was demonstrated that two populations of *P. aeruginosa* exist, one motile subpopulation forming the outer layer of the structure and able to move into the bleached region and one non-motile population forming the core of the microcolony (Fig. 11a–c) [72]. Furthermore, the phenotypic diversity and two-population development were demonstrated using a 1:1 mixture of a wild-type *P. aeruginosa* tagged with YFP and a non-motile *pilA* mutant of *P. aeruginosa* tagged with CFP. The role of the two populations in forming the final biofilm structure was in this way shown on a single-cell level as well as on a 3D structural level (Fig. 11d, e) [80, 103]. Interestingly, it was found that upon antimicrobial treatment of this biofilm, only one of the two subpopulations was sensitive to compounds like colistin,

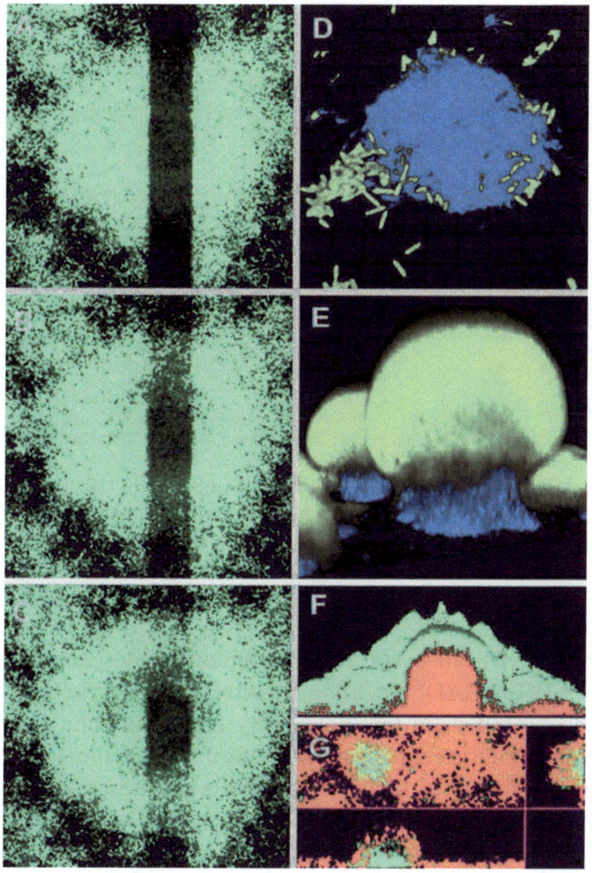

Fig. 11 Micrographs showing a *top-down* view of GFP-tagged PAO1 cells in a microcolony. FRAP was applied to the colony by bleaching a rectangular area across the colony (**a**), followed by time-lapse recording using confocal laser scanning microscopy of the cells. **b** and **c** Show how only cells located in the periphery of the microcolony are motile and cover the bleached area [72]. **d** and **e** The initial development and the final mature 3D structure respectively of a mixture of motile *PAO1* tagged with YFP (*yellow*) and *a PAO1* pilA mutant tagged with CFP (*blue*) [103]. Reproduced with permission from John Wiley & Sons. Micrographs **f**, **g** shows 3D structure representations of live (GFP-tagged *green* cells) and dead [propidium iodide (PI)-stained *red* cells] distributed after treatment with colistin and ciprofloxacin, respectively [72, 102 and Haagensen, unpublished]

tetracycline, ciprofloxacin, etc., implying that *P. aeruginosa* forms different subpopulations to have a higher chance of handling incoming perturbations. The surviving subpopulation of cells exhibits phenotypic tolerance and not resistance, as surviving biofilm cells harvested from antimicrobial-treated biofilms exhibit the same antimicrobial susceptibility phenotype as the cells that were used to initiate the biofilm (Fig. 11f, g) [72, 102].

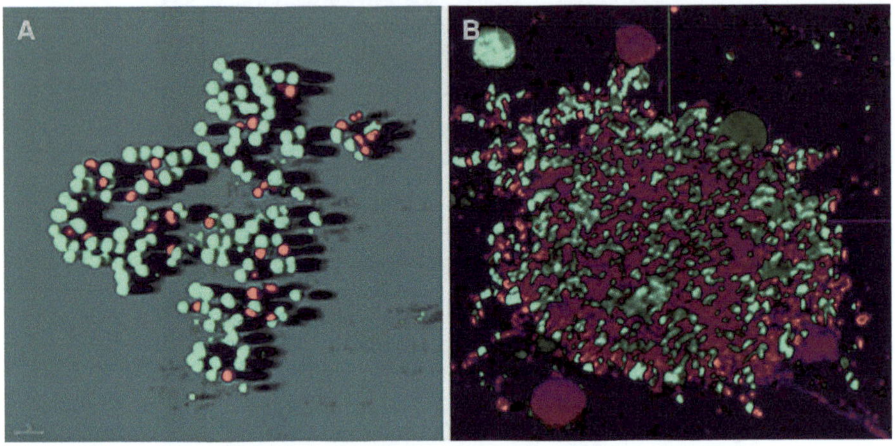

Fig. 12 Micrograph shows 3D structure representations of live (Syto9-stained *green* cells) and dead (PI-stained *red* cells) cells from cystic fibrosis patients undergoing intensive antibiotic treatment. Cells were stained immediately after sampling (Haagensen unpublished)

In sputum obtained from cystic fibrosis patients undergoing intensive antibiotic treatment, LIVE/DEAD® staining (Invitrogen, CA) in combination with confocal microscopy indicated that potentially persistent cells appear frequently. Colistin, a membrane-targeting agent, was used during the treatment, and cell sorting followed by plating of live and dead cells confirmed the existence of two populations and persistence of cells also upon antibiotic used in treatment (Fig. 12). As described for *E. coli*, growth rate-dependent development of tolerant cells and phenotypic differentiation are important survival strategies also for *P. aeruginosa*.

4 Yeast Single-cell and Biofilm Microscopy

Historically, microscopy of yeast cells has focused on free-living cells in pure culture. In this respect the yeast *Saccharomyces cerevisiae* has served as a successful model for the study of organelles and cell structures in the eukaryotic cell. *S. cerevisiae* is easily genetically modified, has a fast reproduction time and perhaps most importantly a large number of community resources exists, such as a complete set of GFP-tagged proteins and targeted knockout mutants of all genes in the genome [45, 71]. Though relatively small in comparison to other eukaryotic cells, with a cell size of 3–5 μm, *S. cerevisiae* live cell imaging has excelled in recent years with the development of STED microscopy and beam-scanning multifocal multiphoton confocal microscopy, which have led to resolution limits much below 100 nm [41, 79] (and Stelzer, this volume).

4.1 Cell-to-Cell Variation in Yeast Populations

As with bacteria, biochemical assays for yeast cells are based on the assumption that protein expression and localization are uniform throughout a population of isogenic cells. Genetic tools however are widely used to investigate cell-to-cell variation within a population. Examples of these are mating type switching in haploid cells, mating between cells of different mating type [129], a morphological shift in *S. cerevisiae* from a yeast to a pseudohyphal form and a shift in *Candida albicans* from a yeast to a filamentous phenotype [46, 88]. More recently, microscopic methods have been applied to screen for cell-to-cell variation in gene expression within populations.

GFP tagging of 4,156 *S. cerevisiae* proteins in individual cell lines has been used to identify and verify protein localization by fluorescence microscopy [71]. While many proteins have specific cellular localization, some proteins vary their localization between cells. A classical example of this is the uneven distribution of the *S. cerevisiae* mating type switching protein (Ash1p) between mother and daughter cells. Ash1p is asymmetrically distributed in a way that the concentration of Ash1p is higher in the daughter nucleus where it inhibits mating type switching [124]. Ash1p asymmetry is regulated at the mRNA level, with mRNA synthesized in the mother cells being transported to the daughter cell. This was elegantly shown in living yeast cells by fluorescence microscopy of *ASH1* mRNAs interacting with a GFP-labeled MS2 bacteriophage coat protein through a stem loop structure introduced into the mRNA [12]. Tagging of mRNA with GFP-MS2 was later used to show that most mRNA species in *S. cerevisiae* have a specific location in the cell and that localization is uniform among cells in a population [53, 151].

Protein abundance seems to vary much within populations. To understand the background of variation, Weissman and coworkers measured the abundance of 2500 protein in individual clones at the single-cell level [101]. They performed high-throughput flow cytometry of a library of GFP-tagged yeast strains and discovered that variation in protein expression is largely caused by stochastic variation at the level of mRNA. Interestingly, there are drastic differences in noise between the functional classes of proteins in *S. cerevisiae*. Genes responding to environmental changes encode gene products with large variation, while proteins involved in structural processes vary less. These differences may reflect selective pressure for a given level of variation, where cellular processes that require accuracy will select for low variation. Large variation, however, may permit a population to express multiple phenotypes to optimize average fitness in changing environments.

4.2 Bistability in Yeasts

Subpopulations with inheritable differential expression of certain genes due to bistability are known from the common human opportunistic pathogenic yeast *C.*

albicans. *C. albicans* can switch between spherical cells that form white colonies and bigger elongated cells that form opaque colonies [110, 125]. Bistability is regulated at the genetic level by the transcription factor Wor1p that is present in very low amounts in white cells and accumulates in opaque cells. Wor1p binds to the promoter of its own gene and induces its expression in a positive feedback loop [69, 152]. The positive feedback loop in combination with stochastic variation in Wor1p expression is suggested to be responsible for switching and inheritance of white and opaque states.

While investigations of bistability in *C. albicans* have been driven by macroscopic features, advanced microscopy such as microscopic high-content screening and high throughput flow cytometry combined with libraries of strains with GFP-tagged proteins may likely lead to the discovery of bistability in *S. cerevisiae* [33, 71, 101, 142].

4.3 Microscopy of Candida albicans *Biofilm*

Microscopy of yeast communities has become increasingly important with a growing number of human infections caused by fungal biofilm on catheters and implants (recently reviewed by Ramage [108]). Biofilms from the most common fungal pathogens, *Candida* spp., are currently being studied by scanning electron microscopy and confocal laser scanning microscopy (CLSM). While scanning electron microscopy reveals biofilm organization and extracellular matrix [26], CLSM can be used to monitor live cell biofilm development in three dimensions over time [26, 108, 120]. Recently atomic force microscopy has also been used to visualize the surface structure of the *C. albicans* biofilm, though the full potential of the AFM for this purpose has probably not been fully exploited yet [84].

A mature biofilm of *C. albicans* is composed of two morphotypes, a unicellular yeast and a multicellular hyphal form [6, 26]. Scanning electron microscopy revealed that both morphotypes can form biofilm individually, though the coexistence of hyphae with yeast cells appears to be essential for a dense biofilm [6]. Deletion of genes essential for filamentous growth in *C. albicans* leads to thin biofilms composed solely of unicellular yeast cells. Biofilms composed of hyphae are dense and appear to lack channel-like structures found in the wild-type *C. albicans* biofilms.

While morphology of *C. albicans* biofilms is well understood, little is known about the molecular mechanism underlying biofilm formation [108]. A limiting factor in this respect is the choice of model organism. Genetic modifications and screens in *Candida* species are cumbersome and often hampered by the existence of paralogous genes that mask the phenotype of a gene deletion. The closely related yeast, *S. cerevisiae*, offers an attractive alternative with its ability to form biofilm [109] and the ease by which genetic modifications can be carried out in this organism. To further develop *S. cerevisiae* as a model for biofilm studies, we have recently developed CLSM methods for the study of biofilms of this organism.

4.4 Confocal Microscopy of S. cerevisiae Biofilms on Batch Culture Slides and in Flow Chambers

Saccharomyces cerevisiae biofilm formation and development can be studied by CLSM (Fig. 13) applying the batch culture slide method described for *C. albicans* and *C. glabrata* with the modifications described in the figure legends to Fig. 13 [120].

With few exceptions, *S. cerevisiae* biofilms can be studied in flow cells according to the protocol applied for *P. aeruginosa* (Fig. 14; [131]). A necessary development was the choice of surface where *S. cerevisiae* can adhere and form biofilm. While *P. aeruginosa* and many other bacteria readily attach to silica surfaces, *S. cerevisiae* adheres poorly to glass surfaces. Suitable surfaces for *S. cerevisiae* biofilm formation are plastics such as, e.g., polyethylene (PE), polypropylene (PP) and to a lesser extent polyvinylchloride (PVC), which were first described for *S. cerevisiae* biofilm assays [109]. Besides these polymers, different polyesters and slides coated with collagen or poly-L-lysine are suitable substrates for *S. cerevisiae* biofilm assays in batch as well as flow cells (unpublished). While plastic cover slides are applicable as biofilm surfaces, several of them suffer from autofluorescence that disturbs CLSM visualization. Coverslips such as the Thermanox PE are autofluorescent in the range 380–545 nm, excluding work with blue and green dyes such as the vital stain Syto9 (Invitrogen. Irvine, CA), GFP and CFP. PVC on the other hand is without autofluorescence in the visible range and therefore an optimal choice as a surface for yeast biofilm imaging (Fig. 15).

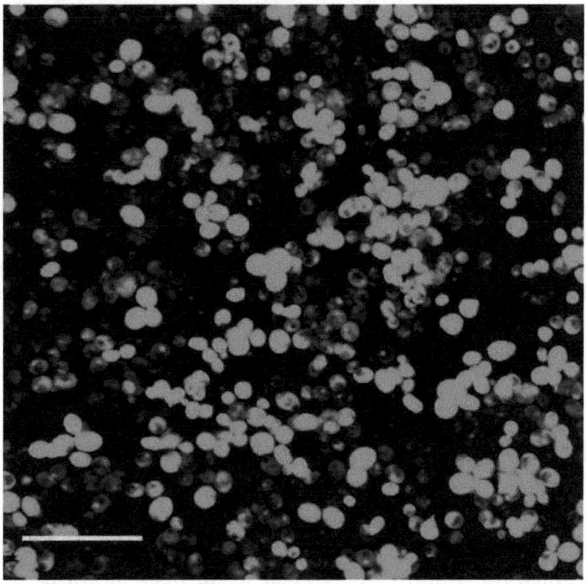

Fig. 13 CLSM of *S. cerevisiae* (CEN.PK113.7D *sfl1*) batch biofilm after 24-h growth in synthetic complete medium with amino acids and 2% glucose (SC-ura). Cells were grown in a Lab-Tek™ Chamber Slide™ System; Permanox® (NUNC, Denmark) in 1 ml medium and stained 30 min with Syto9. *Bottom left bar* 30 μm. CLSM was performed with a Zeiss LSM510 microscope using a 63x/0.95NA water immersion lens. *Bar* 30 μm

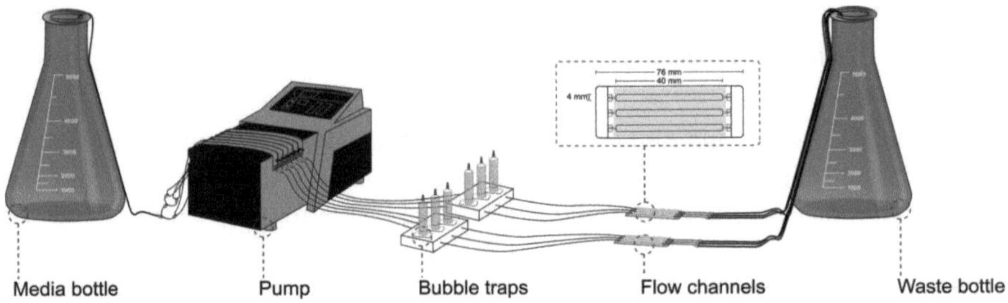

Fig. 14 Experimental setup for *S. cerevisiae* biofilm in flow cells. A flow cell with flow channels is covered with a PVC cover slip that serves as a surface for yeast biofilm attachment and development. A peristaltic pump ensures constant flow of media from the media bottle through the flow channels

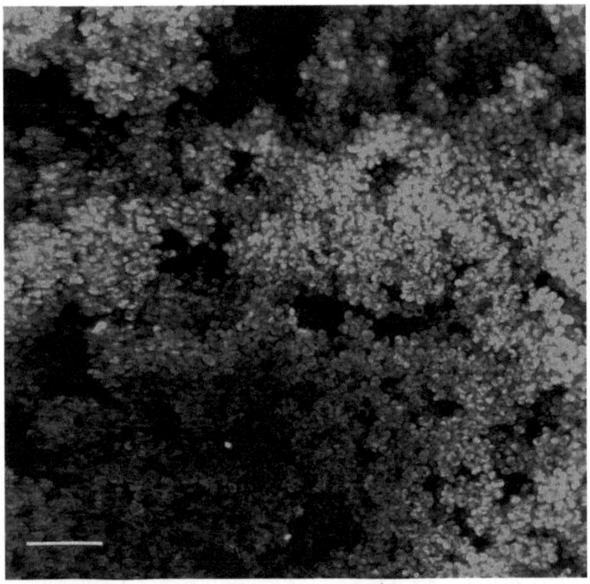

Fig. 15 CLSM of *S. cerevisiae* (CEN.PK113.7D *sfl1*) biofilm in the flow cell setup shown in Fig. 14. Image was recorded after 42 h growth in continuous flow of synthetic complete medium with amino acids and 0.02% glucose (SC-ura). Biofilm formed on PVC cover slips (Rinzl, Electron Microscopy Sciences, Hatfield, PA) and was visualized by staining with Syto9. CLSM was performed with a Zeiss LSM510 microscope using a 40x/1.3NA oil immersion lens. *Bar* 30 μm

Secondly, *S. cerevisiae* biofilm formation is dependent on a haploid cell state [109] and expression of cell surface adhesins such as Flo11p (Muc1p) or Flo1p. Most laboratory strains as well as many natural isolates of *S. cerevisiae* do not express the *FLO11* gene or other adhesion genes [43], and they are therefore not directly applicable for biofilm studies. The *S. cerevisiae* strain background of choice for biofilm studies has so far been strain Σ1278b [109]. However, mutants of other strain backgrounds such as S288c or CEN.PK may be used to study biofilms of *S. cerevisiae* if these strains express the *FLO1* or *FLO11* genes [43].

Besides the use of *S. cerevisiae* for identification of genes involved in yeast biofilm formation, maturation and detachment, *S. cerevisiae* biofilm in both batch and flow cells can effectively be screened for susceptibility to fungicides. Molecular targets for fungicides as well as fungicide resistance mechanisms may be uncovered by the combination of barcode-target yeast mutants and fluorescence microscopic high-content screening.

5 Future Perspectives

While the novel advances in microscopy have proven useful for investigation of microbial communities and microbial single cells, biological discoveries made on the basis of these technologies are only starting to emerge.

One major opportunity is to resolve live objects below the theoretical optical resolution of 200–250 nm with methods such as STED, STORM and SIM. These improved resolution limits enable increased insight into complex spatiotemporal processes such as cell cycle, DNA repair and DNA organization. Higher resolution of fluorescent tagged proteins and RNAs of unknown function that co-localize with others of known function will further aid in the assignment of function to the large group of genes with hitherto unknown role in the cell. Combined with Raman spectroscopy, high-resolution microscopy can also provide information on the molecular basis, localization and structure of extracellular matrix that is otherwise difficult to obtain.

Nanoscale spatial resolution by AFM facilitates visualization of cell surface structures such as extracellular polymers, flagella and pili. At the macromolecular level, the AFM has already proven to provide novel understanding of interactions between macromolecules, e.g., for protein-DNA binding kinetics.

Another aspect of the new technologies is high content automated screening. Automated high-resolution microscopy combined with libraries of, e.g., GFP-tagged proteins or knock-out mutants allows integrated analysis of antibiotic resistance, signal transduction, expression patterns and many other aspects of cell biology and physiology. As mentioned previously, screening of expression patterns is likely to reveal novel examples of bistabilities in microorganisms that are not readily recognized at the macroscopic level. High content automated screening may also find its application in evolution biology where clones of populations that have undergone experimental evolution may be screened for phenotypic differentiation.

Finally, physiological measurements of microbial cells in complex communities, with methods such as confocal Raman microscopy, will provide valuable insight into the physiology of complex microbial communities with respect to cross feeding between species and cell types, local gradients and organization of communities, and internal changes in metabolism. Information about physiological processes in complex microbial communities has a high impact on our basic understanding of microorganisms and applied applications in biotechnology.

Acknowledgments This work was supported by grants from the Carlsberg Foundation and the Lundbeck Foundation to Janus A. J. Haagensen. Birgitte Regenberg was supported by The Danish Council for Independent Research I Natural Sciences (FNU). Claus Sternberg was supported by a grant from the Danish National Advanced Technology Foundation. The authors wish to thank Dr. Ninell P. Mortensen for information about atomic force microscopy, Dr. Rolf W. Berg for information about coherent anti-Stokes Raman spectroscopy and Rune L. Jensen for Fig. 14.

References

1. Abbe E (1873) Beiträge zur Theorie des Mikroskops und der mikroskopischen Wahrnehmung. Arch Mikrosk Anat 9:413–468
2. Abràmoff MD, Magalhães PJ, Ram SJ (2004) Image processing with ImageJ. Biophotonics Int 11:36–41
3. Alfano RR, Shapiro SL (1970) Direct distortion of electronic clouds of rare-gas atoms in intense electric fields. Phys Rev Lett 24:1217
4. Alfano RR, Shapiro SL (1970) Observation of self-phase modulation and small-scale filaments in crystals and glasses. Phys Rev Lett 24:592
5. Amro NA, Kotra LP, Wadu-Mesthrige K et al (2000) High-resolution atomic force microscopy studies of the *Escherichia coli* outer membrane: structural basis for permeability. Langmuir 16:2789–2796
6. Baillie GS, Douglas LJ (1999) Role of dimorphism in the development of *Candida albicans* biofilms. J Med Microbiol 48:671–679
7. Balaban NQ, Merrin J, Chait R et al (2004) Bacterial persistence as a phenotypic switch. Science 305:1622–1625
8. Ball CS (1966) The early history of the compound microscope. Bios 37:51–60
9. Barken KB, Pamp SJ, Yang L et al (2008) Roles of type IV pili, flagellum-mediated motility and extracellular DNA in the formation of mature multicellular structures in *Pseudomonas aeruginosa* biofilms. Environ Microbiol 10:2331–2343
10. Bates M, Huang B, Dempsey GT et al (2007) Multicolor super-resolution imaging with photo-switchable fluorescent probes. Science 317:1749–1753
11. Begley RF, Harvey AB, Byer RL (1974) Coherent anti-Stokes Raman spectroscopy. Appl Phys Lett 25:387–390
12. Bertrand E, Chartrand P, Schaefer M et al (1998) Localization of ASH1 mRNA particles in living yeast. Mol Cell 2:437–445
13. Betzig E, Patterson GH, Sougrat R et al (2006) Imaging intracellular fluorescent proteins at nanometer resolution. Science 313:1642–1645
14. Beyenal H, Donovan C, Lewandowski Z et al (2004) Three-dimensional biofilm structure quantification. J Microbiol Methods 59:395–413
15. Binnig G, Quate CF, Gerber C (1986) Atomic force microscope. Phys Rev Lett 56:930
16. Birk H, Engelhardt J, Storz R et al (2002) Programmable beam-splitter for confocal laser scanning microscopy. In: Three-dimensional and multidimensional microscopy: image acquisition and processing, vol IX. SPIE, San Jose
17. Bolshakova AV, Kiselyova OI, Filonov AS et al (2001) Comparative studies of bacteria with an atomic force microscopy operating in different modes. Ultramicroscopy 86:121–128
18. Borlinghaus R, Gugel H, Albertano P et al (2006) Closing the spectral gap: the transition from fixed-parameter fluorescence to tunable devices in confocal microscopy. In: Three-

dimensional and multidimensional microscopy: image acquisition and processing, vol XIII. SPIE, San Jose
19. Braga PC, Ricci D (1998) Atomic force microscopy: application to investigation of *Escherichia coli* morphology before and after exposure to cefodizime. Antimicrob Agents Chemother 42:18–22
20. Budich C, Neugebauer U, Popp J et al (2008) Cell wall investigations utilizing tip-enhanced Raman scattering. J Microsc 229:533–539
21. Caldwell DE, Korber DR, Lawrence JR (1992) Imaging of bacterial cells by fluorescence exclusion using scanning confocal laser microscopy. J Microbiol Methods 15:249–261
22. Camesano TA, Natan MJ, Logan BE (2000) Observation of changes in bacterial cell morphology using tapping mode atomic force microscopy. Langmuir 16:4563–4572
23. Carreira LA, Goss LP, Malloy TB Jr (1981) Applications of CARS to condensed phase systems. In: Harvey AB (ed) Chemical applications of nonlinear Raman spectroscopy. Academic Press, New York
24. Chai Y, Chu F, Kolter R et al (2008) Bistability and biofilm formation in *Bacillus subtilis*. Mol Microbiol 67:254–263
25. Chan JW, Winhold H, Lane SM et al (2005) Optical trapping and coherent anti-Stokes Raman scattering (CARS) spectroscopy of submicron-size particles. Sel Top Quantum Electron J IEEE 11:858–863
26. Chandra J, Kuhn DM, Mukherjee PK et al (2001) Biofilm formation by the fungal pathogen *Candida albicans*: development, architecture, and drug resistance. J Bacteriol 183: 5385–5394
27. Chatterjee A, Kaznessis YN, Hu WS (2008) Tweaking biological switches through a better understanding of bistability behavior. Curr Opin Biotechnol 19:475–481
28. Christensen BB, Sternberg C, Andersen JB et al (1999) Molecular tools for study of biofilm physiology. Methods Enzymol 310:20–42
29. Costerton JW, Lewandowski Z, DeBeer D et al (1994) Biofilms, the customized microniche. J Bacteriol 176:2137–2142
30. Costerton JW (1995) Overview of microbial biofilms. J Ind Microbiol 15:137–140
31. Costerton JW, Lewandowski Z, Caldwell DE et al (1995) Microbial biofilms. Annu Rev Microbiol 49:711–745
32. Costerton JW (1999) Introduction to biofilm. Int J Antimicrob Agents 11:217–221
33. Czechowska K, Johnson DR, van dM Jr (2008) Use of flow cytometric methods for single-cell analysis in environmental microbiology. Curr Opin Microbiol 11:205–212
34. Daims H, Lucker S, Wagner M (2006) daime, a novel image analysis program for microbial ecology and biofilm research. Environ Microbiol 8:200–213
35. Davidson CJ, Surette MG (2008) Individuality in bacteria. Annu Rev Genet 42:253–268
36. Denk W, Strickler JH, Webb WW (1990) Two-photon laser scanning fluorescence microscopy. Science 248:73–76
37. Dubnau D (1991) The regulation of genetic competence in *Bacillus subtilis*. Mol Microbiol 5:11–18
38. Dubnau D (1991) Genetic competence in *Bacillus subtilis*. Microbiol Rev 55:395–424
39. Dubnau D, Losick R (2006) Bistability in bacteria. Mol Microbiol 61:564–572
40. Dyba M, Keller J, Hell SW (2005) Phase filter enhanced STED-4pi fluorescence microscopy: theory and experiment. New J Phys 134:1–21
41. Egner A, Jakobs S, Hell SW (2002) Fast 100-nm resolution three-dimensional microscope reveals structural plasticity of mitochondria in live yeast. Proc Natl Acad Sci USA 99:3370–3375
42. Eklund H, Roos A, Eng ST (1975) Rotation of laser beam polarization in acousto-optic devices. Opt Quant Electron 7:73–79
43. Fichtner L, Schulze F, Braus GH (2007) Differential Flo8p-dependent regulation of FLO1 and FLO11 for cell-cell and cell-substrate adherence of *S. cerevisiae* S288c. Mol Microbiol 66:1276–1289

44. Gad M, Ikai A (1995) Method for immobilizing microbial cells on gel surface for dynamic AFM studies. Biophys J 69:2226–2233
45. Giaever G, Chu AM, Ni L et al (2002) Functional profiling of the *Saccharomyces cerevisiae* genome. Nature 418:387–391
46. Gimeno CJ, Ljungdahl PO, Styles CA et al (1992) Unipolar cell divisions in the yeast *Saccharomyces-Cerevisiae* lead to filamentous growth—regulation by starvation and Ras. Cell 68:1077–1090
47. Gonzalez-Pastor JE, Hobbs EC, Losick R (2003) Cannibalism by sporulating bacteria. Science 301:510–513
48. Gustafsson MGL (2000) Surpassing the lateral resolution limit by a factor of two using structured illumination microscopy. J Microsc 198:82–87
49. Gustafsson MGL (2005) Nonlinear structured-illumination microscopy: wide-field fluorescence imaging with theoretically unlimited resolution. Proc Natl Acad Sci USA 102:13081–13086
50. Guthold M, Bezanilla M, Erie DA et al (1994) Following the assembly of RNA polymerase–DNA complexes in aqueous solutions with the scanning force microscope. Proc Natl Acad Sci USA 91:12927–12931
51. Göppert-Mayer M (1931) Über Elementarakte mit zwei Quantensprüngen. Ann Phys 401:273–294
52. Hahn J, Kong L, Dubnau D (1994) The regulation of competence transcription factor synthesis constitutes a critical control point in the regulation of competence in *Bacillus subtilis*. J Bacteriol 176:5753–5761
53. Haim L, Zipor G, Aronov S et al (2007) A genomic integration method to visualize localization of endogenous mRNAs in living yeast. Nat Methods 4:409–412
54. Hall-Stoodley L, Costerton JW, Stoodley P (2004) Bacterial biofilms: from the natural environment to infectious diseases. Nat Rev Microbiol 2:95–108
55. Hallett P, Offer G, Miles MJ (1995) Atomic-force microscopy of the myosin molecule. Biophys J 68:1604–1606
56. Hansma HG, Vesenka J, Siegerist C et al (1992) Reproducible imaging and dissection of plasmid DNA under liquid with the atomic force microscope. Science 256:1180–1184
57. Harke B, Keller J, Ullal CK et al (2008) Resolution scaling in STED microscopy. Opt Express 16:4154–4162
58. Harz M, Rösch P, Peschke KD et al (2005) Micro-Raman spectroscopic identification of bacterial cells of the genus *Staphylococcus* and dependence on their cultivation conditions. Analyst 130:1543–1550
59. Harz M, Rösch P, Popp J (2009) Vibrational spectroscopy—a powerful tool for the rapid identification of microbial cells at the single-cell level. Cytometry Part A 75A:104–113
60. Heimstädt O (1911) Das Fluoreszenzmikroskop. Z Wiss Mikrosk 28:330–337
61. Hein B, Willig KI, Hell SW (2008) Stimulated emission depletion (STED) nanoscopy of a fluorescent protein-labeled organelle inside a living cell. Proc Natl Acad Sci USA 105:14271–14276
62. Heintzmann R, Jovin TM, Cremer C (2002) Saturated patterned excitation microscopy? A concept for optical resolution improvement. J Opt Soc Am A 19:1599–1609
63. Hell SW (2003) Toward fluorescence nanoscopy. Nat Biotechnol 21:1347–1355
64. Hess ST, Girirajan TPK, Mason MD (2006) Ultra-high resolution imaging by fluorescence photoactivation localization microscopy. Biophys J 91:4258–4272
65. Heydorn A, Nielsen AT, Hentzer M et al (2000) Quantification of biofilm structures by the novel computer program COMSTAT. Microbiol 146(10):2395–2407
66. Hirvonen L, Wicker K, Mandula O et al (2009) Structured illumination microscopy of a living cell. Eur Biophys J 38:807–812
67. Hopt A, Neher E (2001) Highly nonlinear photodamage in two-photon fluorescence microscopy. Biophys J 80:2029–2036
68. Huang B, Wang W, Bates M et al (2008) Three-dimensional super-resolution imaging by stochastic optical reconstruction microscopy. Science 319:810–813

69. Huang G, Wang H, Chou S et al (2006) Bistable expression of WOR1, a master regulator of white-opaque switching in *Candida albicans*. Proc Natl Acad Sci USA 103:12813–12818
70. Huang W, Ude S, Spiers A (2007) *Pseudomonas fluorescens* SBW25 biofilm and planktonic cells have differentiable Raman spectral profiles. Microb Ecol 53:471–474
71. Huh WK, Falvo JV, Gerke LC et al (2003) Global analysis of protein localization in budding yeast. Nature 425:686–691
72. Haagensen JA, Klausen M, Ernst RK et al (2007) Differentiation and distribution of colistin- and sodium dodecyl sulfate-tolerant cells in *Pseudomonas aeruginosa* biofilms. J Bacteriol 189:28–37
73. Jarvis RM, Brooker A, Goodacre R (2006) Surface-enhanced Raman scattering for the rapid discrimination of bacteria. Faraday Discuss 132:281–292
74. Jarvis RM, Goodacre R (2008) Characterisation and identification of bacteria using SERS. Chemi Soc Rev 37:931–936
75. Jeanmaire DL, Van Duyne RP (1977) Surface Raman spectroelectrochemistry: Part I. Heterocyclic, aromatic, and aliphatic amines adsorbed on the anodized silver electrode. J Electroanal Chem 84:1–20
76. Juette MF, Gould TJ, Lessard MD et al (2008) Three-dimensional sub-100 nm resolution fluorescence microscopy of thick samples. Nat Methods 5:527–529
77. Kasas S, Ikai A (1995) A method for anchoring round shaped cells for atomic force microscope imaging. Biophys J 68:1678–1680
78. Kearns DB, Losick R (2005) Cell population heterogeneity during growth of *Bacillus subtilis*. Genes Dev 19:3083–3094
79. Klar TA, Jakobs S, Dyba M et al (2000) Fluorescence microscopy with diffraction resolution barrier broken by stimulated emission. Proc Natl Acad Sci USA 97:8206–8210
80. Klausen M, Aaes-Jorgensen A, Molin S et al (2003) Involvement of bacterial migration in the development of complex multicellular structures in *Pseudomonas aeruginosa* biofilms. Mol Microbiol 50:61–68
81. Knight JC, Birks TA, Russell PSJ et al (1996) All-silica single-mode optical fiber with photonic crystal cladding. Opt Lett 21:1547–1549
82. Kolter R, Greenberg EP (2006) Microbial sciences: the superficial life of microbes. Nature 441:300–302
83. König K (2000) Multiphoton microscopy in life sciences. J Microsc 200:83–104
84. Lal P, Sharma D, Pruthi P et al (2009) Exopolysaccharide analysis of biofilm-forming *Candida albicans*. J Appl Microbiol 109:128–136
85. Lawrence JR, Korber DR, Hoyle BD et al (1991) Optical sectioning of microbial biofilms. J Bacteriol 173:6558–6567
86. Le CE, Frechon D, Barray M et al (1994) Observation of binding and polymerization of Fur repressor onto operator-containing DNA with electron and atomic force microscopes. Proc Natl Acad Sci USA 91:11816–11820
87. Lewandowski Z, Beyenal H (2007) Fundamentals of biofilm research. CRC Press, Boca Raton
88. Liu H (2001) Transcriptional control of dimorphism in *Candida albicans*. Curr Opin Microbiol 4:728–735
89. Minsky M (1988) Memoir on inventing the confocal scanning microscope. Scanning 10:128–138
90. Müller M, Zumbusch A (2007) Coherent anti-Stokes Raman scattering microscopy. Chem Phys Chem 8:2156–2170
91. Müller S, Nebe-von-Caron G (2010) Functional single-cell analyses: flow cytometry and cell sorting of microbial populations and communities. FEMS Microbiol Rev 34:554–587
92. Maamar H, Dubnau D (2005) Bistability in the *Bacillus subtilis* K-state (competence) system requires a positive feedback loop. Mol Microbiol 56:615–624
93. Maamar H, Raj A, Dubnau D (2007) Noise in gene expression determines cell fate in *Bacillus subtilis*. Science 317:526–529

94. Naumann D, Keller S, Helm D et al (1995) FT-IR spectroscopy and FT-Raman spectroscopy are powerful analytical tools for the non-invasive characterization of intact microbial cells. J Mol Struct 347:399–405
95. Neu TR, Walczysko P, Lawrence JR (2004) Two-photon imaging for studying the microbial ecology of biofilm systems. Microbes Environ 19:1–6
96. Neu TR, Lawrence JR (2005) One-photon versus two-photon laser scanning microscopy and digital image analysis of microbial biofilms. Methods Microbiol 34:89–136
97. Neu TR, Manz B, Volke F et al (2010) Advanced imaging techniques for assessment of structure, composition and function in biofilm systems. FEMS Microbiol Ecol 72:1–21
98. Neugebauer U, Rösch P, Schmitt M et al (2006) On the way to nanometer-sized information of the bacterial surface by tip-enhanced Raman spectroscopy. Chem Phys Chem 7:1428–1430
99. Neugebauer U, Schmid U, Baumann K et al (2006) Characterization of bacterial growth and the influence of antibiotics by means of UV resonance Raman spectroscopy. Biopolymers 82:306–311
100. Neugebauer U, Schmid U, Baumann K et al (2007) Towards a detailed understanding of bacterial metabolism—spectroscopic characterization of *Staphylococcus epidermidis*. Chem Phys Chem 8:124–137
101. Newman JR, Ghaemmaghami S, Ihmels J et al (2006) Single-cell proteomic analysis of *S. cerevisiae* reveals the architecture of biological noise. Nature 441:840–846
102. Pamp SJ, Gjermansen M, Johansen HK et al (2008) Tolerance to the antimicrobial peptide colistin in *Pseudomonas aeruginosa* biofilms is linked to metabolically active cells, and depends on the pmr and mexAB-oprM genes. Mol Microbiol 68:223–240
103. Pamp SJ, Sternberg C, Tolker-Nielsen T (2009) Insight into the microbial multicellular lifestyle via flow-cell technology and confocal microscopy. Cytometry Part A 75A: 90–103
104. Patterson GH, Betzig E, Lippincott-Schwartz J et al (2007) Devloping Photoactivated Localization Microscopy (PALM). In: 4th IEEE international symposium on biomedical imaging: from nano to macro, ISBI 2007
105. Pavani SRP, Thompson MA, Biteen JS et al (2009) Three-dimensional, single-molecule fluorescence imaging beyond the diffraction limit by using a double-helix point spread function. Proc Natl Acad Sci USA 106:2995–2999
106. Pätzold R, Keuntje M, Anders-von Ahlften A (2006) A new approach to non-destructive analysis of biofilms by confocal Raman microscopy. Anal Bioanal Chem 386:286–292
107. Radmacher M, Fritz M, Hansma HG et al (1994) Direct observation of enzyme-activity with the atomic-force microscope. Science 265:1577–1579
108. Ramage G, Mowat E, Jones B et al (2009) Our current understanding of fungal biofilms. Crit Rev Microbiol 35:340–355
109. Reynolds TB, Fink GR (2001) Bakers' yeast, a model for fungal biofilm formation. Science 291:878–881
110. Rikkerink EH, Magee BB, Magee PT (1988) Opaque-white phenotype transition: a programmed morphological transition in *Candida albicans*. J Bacteriol 170:895–899
111. Rittweger E, Han KY, Irvine SE et al (2009) STED microscopy reveals crystal colour centres with nanometric resolution. Nat Photon 3:144–147
112. Robichon D, Girard J-C, Cenatiempo Y et al (1999) Atomic force microscopy imaging of dried or living bacteria. Comptes Rendus de l'Académie des Sciences - Series III - Sciences de la Vie 322:687–693
113. Rosch P, Harz M, Schmitt M et al (2005) Chemotaxonomic identification of single bacteria by micro-Raman spectroscopy: application to clean-room-relevant biological contaminations. Appl Environ Microbiol 71:1626–1637
114. Ruska E (1987) The development of the electron microscope and of electron microscopy. Rev Mod Phys 59:627
115. Rust MJ, Bates M, Zhuang X (2006) Sub-diffraction-limit imaging by stochastic optical reconstruction microscopy (STORM). Nat Methods 3:793–796

116. Sandt C, Smith Palmer T, Pink J et al (2008) Quantification of local water and biomass in wild type PA01 biofilms by confocal Raman microspectroscopy. J Microbiol Methods 75:148–152
117. Schabert FA, Engel A (1994) Reproducible acquisition of *Escherichia coli* porin surface topographs by atomic force microscopy. Biophys J 67:2394–2403
118. Schermelleh L, Carlton PM, Haase S et al (2008) Subdiffraction multicolor imaging of the nuclear periphery with 3D structured illumination microscopy. Science 320:1332–1336
119. Schuster KC, Urlaub E, Gapes JR (2000) Single-cell analysis of bacteria by Raman microscopy: spectral information on the chemical composition of cells and on the heterogeneity in a culture. J Microbiol Methods 42:29–38
120. Seneviratne CJ, Silva WJ, Jin LJ et al (2009) Architectural analysis, viability assessment and growth kinetics of *Candida albicans* and *Candida glabrata* biofilms. Arch Oral Biol 54:1052–1060
121. Shi L, Günther S, Hübschmann T et al (2007) Limits of propidium iodide as a cell viability indicator for environmental bacteria. Cytometry Part A 71A: 592–598
122. Shroff H, Galbraith CG, Galbraith JA et al (2007) Dual-color superresolution imaging of genetically expressed probes within individual adhesion complexes. Proc Natl Acad Sci USA 104:20308–20313
123. Shtengel G, Galbraith JA, Galbraith CG et al (2009) Interferometric fluorescent super-resolution microscopy resolves 3D cellular ultrastructure. Proc Natl Acad Sci USA 106:3125–3130
124. Sil A, Herskowitz I (1996) Identification of asymmetrically localized determinant, Ash1p, required for lineage-specific transcription of the yeast HO gene. Cell 84:711–722
125. Slutsky B, Staebell M, Anderson J et al (1987) "White-opaque transition": a second high-frequency switching system in *Candida albicans*. J Bacteriol 169:189–197
126. Smith E, Dent G (2005) Modern Raman spectroscopy: a practical approach. Wiley, New York
127. Smits WK, Eschevins CC, Susanna KA et al (2005) Stripping Bacillus: ComK auto-stimulation is responsible for the bistable response in competence development. Mol Microbiol 56:604–614
128. Smits WK, Kuipers OP, Veening J-W (2006) Phenotypic variation in bacteria: the role of feedback regulation. Nat Rev Microbiol 4:259–271
129. Sprague GF Jr, Blair LC, Thorner J (1983) Cell interactions and regulation of cell type in the yeast *Saccharomyces cerevisiae*. Annu Rev Microbiol 37:623–660
130. Sternberg C, Christensen BB, Johansen T et al (1999) Distribution of bacterial growth activity in flow-chamber biofilms. Appl Environ Microbiol 65:4108–4117
131. Sternberg C, Tolker-Nielsen T (2006) Growing and analyzing biofilms in flow cells. Curr Protoc Microbiol Chapter 1:Unit 1B.2
132. Stoodley P, Sauer K, Davies DG et al (2002) Biofilms as complex differentiated communities. Annu Rev Microbiol 56:187–209
133. Thundat T, Allison DP, Warmack RJ et al (1992) Atomic force microscopy of DNA on mica and chemically modified mica. Scanning Microsc 6:911–918
134. Thundat T, Allison DP, Warmack RJ et al (1992) Imaging isolated strands of dna-molecules by atomic force microscopy. Ultramicroscopy 42:1101–1106
135. Tolker-Nielsen T, Molin S (2000) Spatial organization of microbial biofilm communities. Microb Ecol 40:75–84
136. van Sinderen D, Luttinger A, Kong L et al (1995) comK encodes the competence transcription factor, the key regulatory protein for competence development in *Bacillus subtilis*. Mol Microbiol 15:455–462
137. Vaziri A, Tang J, Shroff H et al (2008) Multilayer three-dimensional super resolution imaging of thick biological samples. Proc Natl Acad Sci USA 105:20221–20226
138. Veening J-W, Smits WK, Hamoen LW et al (2006) Single cell analysis of gene expression patterns of competence development and initiation of sporulation in *Bacillus subtilis* grown on chemically defined media. J Appl Microbiol 101:531–541

139. Veening J-W, Stewart EJ, Berngruber TW et al (2008) Bet-hedging and epigenetic inheritance in bacterial cell development. Proc Natl Acad Sci USA 105:4393–4398
140. Veening JW, Smits WK, Hamoen LW et al (2004) Visualization of differential gene expression by improved cyan fluorescent protein and yellow fluorescent protein production in *Bacillus subtilis*. Appl Environ Microbiol 70:6809–6815
141. Veening JW, Stewart EJ, Berngruber TW et al (2008) Bet-hedging and epigenetic inheritance in bacterial cell development. Proc Nat lAcad Sci USA 105:4393–4398
142. Vizeacoumar FJ, van Dyk N, Vizeacoumar S et al (2010) Integrating high-throughput genetic interaction mapping and high-content screening to explore yeast spindle morphogenesis. J Cell Biol 188:69–81
143. Vlamakis H, Aguilar C, Losick R et al (2008) Control of cell fate by the formation of an architecturally complex bacterial community. Genes Dev 22:945–953
144. Wagner M, Ivleva NP, Haisch C et al (2009) Combined use of confocal laser scanning microscopy (CLSM) and Raman microscopy (RM): investigations on EPS-matrix. Water Res 43:63–76
145. Wildanger D, Rittweger E, Kastrup L et al (2008) STED microscopy with a supercontinuum laser source. Opt Express 16:9614–9621
146. Willig KI, Kellner RR, Medda R et al (2006) Nanoscale resolution in GFP-based microscopy. Nat Methods 3:721–723
147. Willig KI, Harke B, Medda R et al (2007) STED microscopy with continuous wave beams. Nat Methods 4:915–918
148. Yang L, Liu Y, Sternberg C et al (2010) Evaluation of enoyl-acyl carrier protein reductase inhibitors as *Pseudomonas aeruginosa* quorum-quenching reagents. Molecules 15:780–792
149. Yang X, Beyenal H, Harkin G et al (2000) Quantifying biofilm structure using image analysis. J Microbiol Methods 39:109–119
150. Zernike F (1935) Das Phasenkontrastverfahren bei der mikroskopischen Beobachtung. Z Tech Phys 16:454–457
151. Zipor G, Haim-Vilmovsky L, Gelin-Licht R et al (2009) Localization of mRNAs coding for peroxisomal proteins in the yeast, *Saccharomyces cerevisiae*. Proc Natl Acad Sci USA 106:19848–19853
152. Zordan RE, Galgoczy DJ, Johnson AD (2006) Epigenetic properties of white-opaque switching in *Candida albicans* are based on a self-sustaining transcriptional feedback loop. Proc Natl Acad Sci USA 103:12807–12812
153. Zumbusch A, Holtom GR, Xie XS (1999) Three-dimensional vibrational imaging by coherent anti-Stokes Raman scattering. Phys Rev Lett 82:4142

Adv Biochem Engin/Biotechnol (2011) 124: 55–81
DOI: 10.1007/10_2010_97

Published Online: 12 November 2010

Algebraic and Geometric Understanding of Cells: Epigenetic Inheritance of Phenotypes Between Generations

Kenji Yasuda

Abstract We have developed methods and systems for analyzing epigenetic information in cells, as well as that of genetic information, to expand our understanding of how living systems are determined. Because cells are minimum units reflecting epigenetic information, which is considered to map the history of a parallel-processing recurrent network of biochemical reactions, their behaviors cannot be explained by considering only conventional DNA information-processing events. The role of epigenetic information in cells, which complements their genetic information, was inferred by comparing predictions from genetic information with cell behavior observed under conditions chosen to reveal adaptation processes and community effects. A system for analyzing epigenetic information was developed starting from the twin complementary viewpoints of cell regulation as an 'algebraic' system (emphasis on temporal aspects) and as a 'geometric' system (emphasis on spatial aspects). The knowledge acquired from this study may lead to the use of cells that fully control practical applications like cell-based drug screening and the regeneration of organs.

Keywords On-chip single-cell-based cultivation/analysis · Epigenetic information · Algebraic viewpoint · Geometric viewpoint · Individuality

Contents

K. Yasuda (✉)
Department of Biomedical Information, Institute of Biomaterials and Bioengineering, Tokyo Medical and Dental University, 2-3-10 Kanda-Surugadai, Chiyoda-ku, Tokyo 101-0062, Japan
e-mail: yasuda.bmi@tmd.ac.jp

1 Introduction: On-Chip Cultivation Methods for 'Algebraic' and 'Geometric' Viewpoints

Cells are minimum units determining their responses through genetic and epigenetic information like the history of interactions between them and fluctuations in environmental conditions affecting them. The cells in a group are also individual entities, and their differences arise even among cells with identical genetic information that have grown under the same conditions. These cells respond differently to perturbations [1]. Why and how do these differences arise? To understand the rules underlying possible differences occurring in cells, we need to develop methods of simultaneously evaluating both the genetic and epigenetic information not only for molecular level measurement but also for functional measurement. In other words, if we are to understand topics like variations in cells with the same genetic information, inheritance of non-genetic information between adjacent generations of cells, cellular adaptation processes caused by environmental change, the community effect of cells, we also need to analyze their epigenetic information. We thus started a series of studies to analyze epigenetic information among neighboring generations of cells and in the spatial structures of cell network to expand our understanding of how the fates of living systems are determined. As cells are minimum units reflecting epigenetic information, which is considered to map the history of a parallel-processing recurrent network of biochemical reactions, their behaviors cannot be explained by considering only conventional DNA information-processing events. The role of epigenetic information in the higher complexity of cellular groups, which complements their genetic information, is inferred by comparing predictions from genetic information with cell behavior observed under conditions chosen to reveal adaptation processes and community effects. A system for analyzing epigenetic information should be developed starting from the twin complementary viewpoints of cell regulation as an 'algebraic' system (emphasis on temporal aspects; adaptation among generations) and as a 'geometric' system (emphasis on spatial aspects; spatial pattern-dependent community effect). The acquired knowledge should lead us not only to understand the mechanism of the inheritable epigenetic memory but also to be able to control the epigenetic information by the designed sequence of the external stimulation.

As we can see in Fig. 1, the strategy behind our on-chip microfabrication method is constructive, involving three steps. First, we purify cells from tissue one

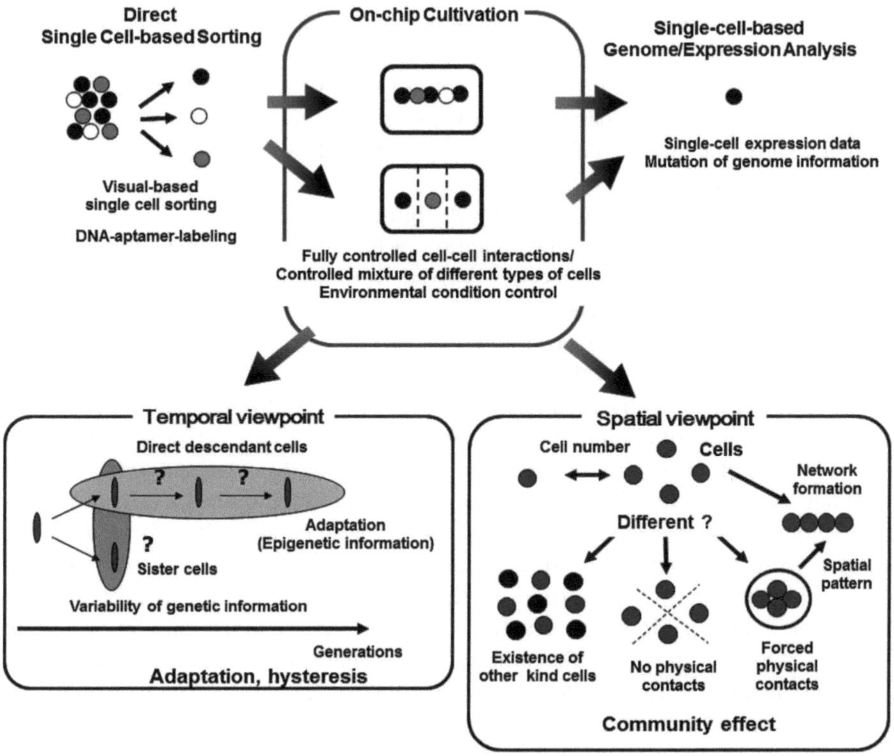

Fig. 1 Our strategy: on-chip single-cell-based analysis with the aim of probing temporal and spatial aspects of cell regulation

by one in a nondestructive manner such as using ultrahigh-speed camera-based real-time cell sorting, or digestible DNA-aptamer labeling [2]. We then cultivate and observe them under fully controlled conditions (e.g., cell population, network patterns, or nutrient conditions) using an on-chip single-cell cultivation chip [3–12] or an on-chip agarose microchamber system [13–20]. Finally, we perform single-cell-based genome/proteome analysis through photothermal denaturation and single-molecule level analysis [21].

In this chapter, we explain the aims of our single-cell-based study using the single-cell-based cultivation/analysis system and introduce some of the results focusing on the 'algebraic' understanding of cellular systems using *Escherichiacoli* cells.

2 Cultivation System for 'Algebraic' Viewpoint: On-Chip Single-Cell Cultivation System for Isolated *E. coli* Cells

The first aim of our single-cell-based study was to develop methods and systems using on-chip microcultivation system that enable the mechanism responsible for controlling (regulating) cells epigenetically to be analyzed for 'algebraic' understanding. The advantage of this approach is that it removes the complexity in underlying physicochemical reactions that are not always completely understood and for which most of the necessary variables cannot be measured. Moreover, this approach shifts the view of cell regulatory processes from a basic chemical ground to a paradigm of the cell as an information-processing unit working as an intelligent machine capable of adapting to changing environmental and internal conditions. This is an alternative representation of the cell and can bring new insights into cellular processes. Thus, models derived from such a viewpoint can directly help in more traditional biochemical and molecular biological analyses that assist in our understanding of control in cells.

Phenotypic and behavioral variations from cell to cell have been observed to exist even in a genetically identical population [1, 22–25]. The resulting heterogeneity in a clonal population may well be important not only for survival [24], but also for cooperation in a population that must obviously exist and work in multicellular organisms [26–28]. The mechanisms of producing phenotypic variations are explored both theoretically [29–34] and experimentally [35–37] as an intracellular noise (fluctuation, or stochastic transcription/translation)-driven process [38]. McAdams and Arkin proposed that stochasticity in the process of gene expression could lead to the substantially large difference in the amount of protein products, which eventually affects the switching mechanisms in individual cells in a group that select between alternative phenotypes [33]. The existence of noise in the gene expression process was shown experimentally by van Oudenaarden and colleagues [37]. They showed that the resulting phenotypic noise had a strong positive correlation with translational efficiency, in contrast to the weak positive correlation observed for transcriptional efficiency. As another example of the experiment, Elowitz and colleagues [36] examined the contributions to overall variation from the gene expression process and from other cellular components separately, showing that the noise in the gene expression process did not uniquely determine the total variability.

These studies are based on the temporal observation of a cell group. The group-based observation, however, cannot show how an individual cell produces different phenotypes and behaviors in the course of proliferation and whether phenotypes and behaviors specific to an individual cell can be inherited. Flow cytometry enables us to obtain the distributions of parameters like concentration, size, shape, DNA content etc. at the single-cell level in a group and is a powerful method to check huge numbers of cells within a short time [39]. To confirm the acquired results from flow cytometry, as a complementary supporting approach, it is desirable to acquire the continuous trackings of a specific single cell's dynamics under the specific circumstances like isolated conditions or comparison of neighboring generations of single cells. In other words, cytometry can give us information about the average properties of huge numbers of cells, i.e., how the group changes, but it cannot give us information about how a single cell changes. Direct microscopic measurement of cells in solid media like cultivation plates [28, 40–43] can identify individual cells and thus can track specific cells continuously.

Although we can begin cultivating cells under isolated conditions and establish the desired connections by controlling the initial spread concentration, it is impossible to keep cells isolated or track more than ten direct descendant generations especially after cell divisions have occurred, and it is impossible to control the interactions between particular cells because the positions of the cells are fixed at the beginning of the cultivation. Thus this on-chip measurement system is thought to be complementary to these conventional methods for gaining an understanding of single-cell level interactions of particular cells.

New techniques are needed to clarify the interactions between genetically identical cells, and for this purpose, we have developed an on-chip single-cell-based microculture method exploiting recent microfabrication techniques and conventional in vivo techniques. To manipulate cells in microchambers, we use non-contact forces, such as optical tweezers and acoustic radiation force, which have been used to handle cells, organelles, and biomolecules on microscope specimens [2, 44–49].

2.1 On-Chip Single-Cell Cultivation System Design

To directly compare sister or direct descendant cells, we developed an on-chip single-cell cultivation system. It enabled excess cells to be transferred from the analysis chamber to the waste chamber through a narrow channel that allowed a particular cell to be selected from cells in the microfabricated cultivation chamber with non-contact force, optical tweezers (Fig. 2).

Figure 3 is a schematic drawing of the procedure of isolation of single cells using optical tweezers, and the entire system we used for on-chip single-cell-based

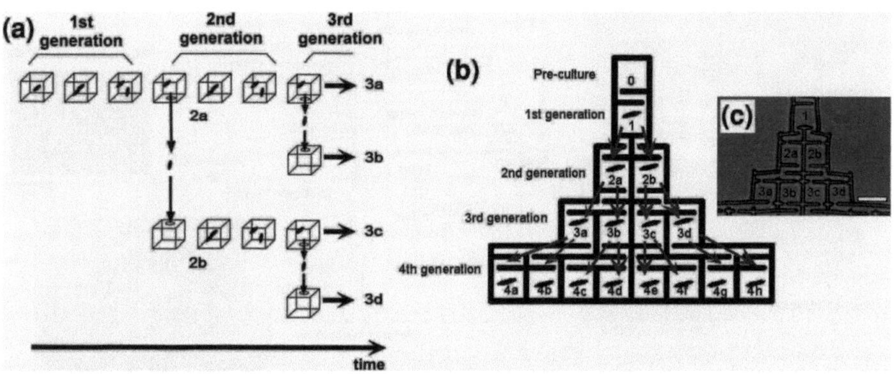

Fig. 2 Single-cell cultivation in microchambers to measure variability in genetic information. **a** Schematic drawing of concept of on-chip single cell cultivation using microchamber system, **b** an example of microchamber array design for single cell cultivation for four generations, **c** micrograph of microchamber array structure made of polydimethylsiloxane (PDMS) on the glass slide, *bar* 50 μm

analysis. It consists of a microchamber array plate (chip), a cover chamber attached to the medium circulation unit, a ×100 phase-contrast/fluorescent microscope, and optical tweezers. The microchamber arrays are the microfabricated structure etched on the glass slide (Fig. 4a–e), or made of thick photo-resist SU-8-5 on the glass slide (Microlithography Chemical Corp., MA) (Fig. 4f). The height of the microchamber array is at least 5 μm, in which the cells are enclosed. The microchamber array is sealed with a semipermeable membrane to prevent the cells escaping from it. The semipermeable membrane is decorated with avidin and the glass slide with biotin to ensure the seal is tight (Fig. 5). With these decorations on the membrane and slide, it is possible to observe cells in the microchamber without them escaping. The microchamber is composed of two main parts and the first is the observation area, which has four compartments in it at the center of the microchamber. Each compartment has a volume of 20 × 20 × 5 μm. Each compartment has four observation sub-compartments (A, B, C, and D in Fig. 4f) at the center of the microstructure. The second part includes the discarding areas at both sides of the microchamber. These enclose surplus cells in observations. The first four direct descendant cells derived from a single cell were placed in one of the four compartments individually to keep them isolated. The excess descendant cells were transferred to the two large compartments (discarding areas) along the white arrow with the optical tweezers. Cells were transferred from the observation area to the discarding area by using the optical tweezers through the narrow path along the white arrow in Fig. 4f. As we can see in the micrographs, only one cell is enclosed in each of the four compartments of the observation area

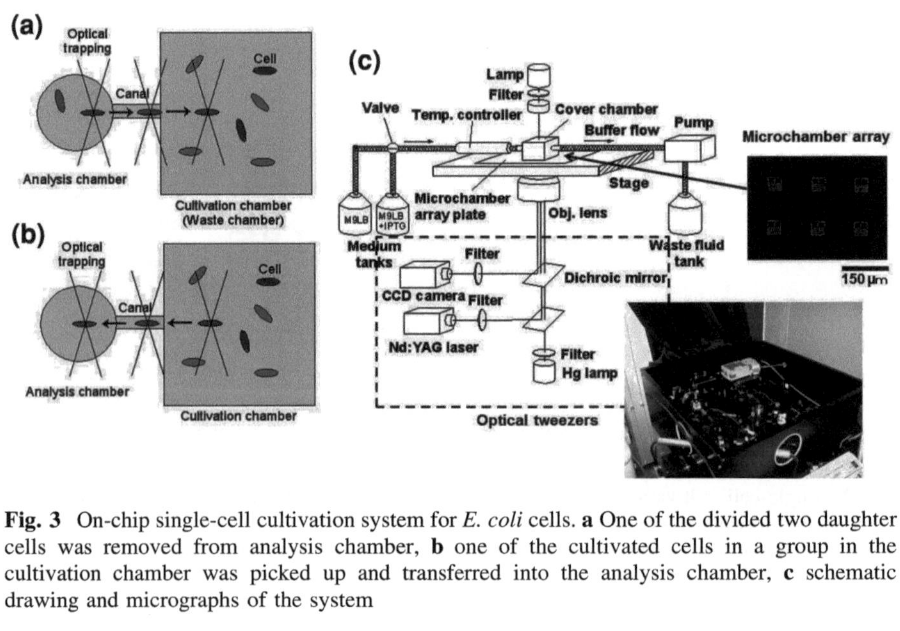

Fig. 3 On-chip single-cell cultivation system for *E. coli* cells. **a** One of the divided two daughter cells was removed from analysis chamber, **b** one of the cultivated cells in a group in the cultivation chamber was picked up and transferred into the analysis chamber, **c** schematic drawing and micrographs of the system

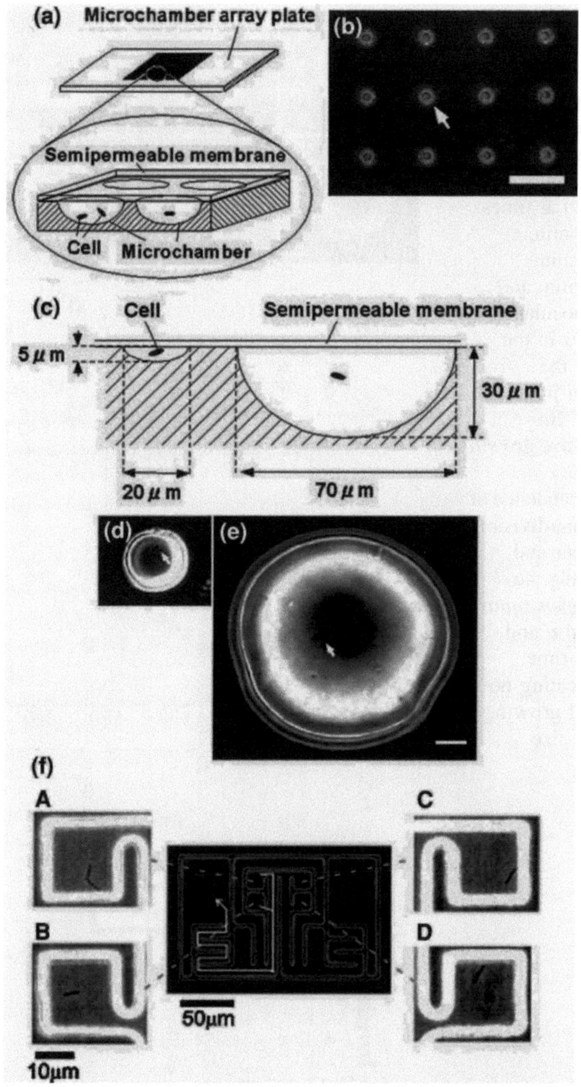

Fig. 4 Optical micrographs of microchambers. **a** Schematic drawing of the microchamber array plate. An $n \times n$ ($n = 20$–50) array of microchambers is etched into a 0.17-mm-thick glass slide. Each microchamber is covered with a semipermeable membrane separating the chamber from the nutrient medium circulating through the medium bath. A single cell or group of cells in the microchamber can thus be isolated from others perfused with the same medium. **b** Optical micrograph of the microchamber array plate. The *arrow* indicates the position of one microchamber with a 20-μm diameter and 5-μm depth. *Bar* 100 μm. **c** Schematic drawing of the two sizes of microchambers. The 'small chamber' on the *left* has a 20-μm diameter and 5-μm depth (shown in **b**), and the 'large chamber' on the *right* has a 70-μm diameter and 30-μm depth. Optical micrographs of **d** a small chamber, and **e** a large chamber. The *arrows* indicate the positions of *E. coli* in the chambers. *Bar* 10 μm. **f** Microchamber array with four compartments

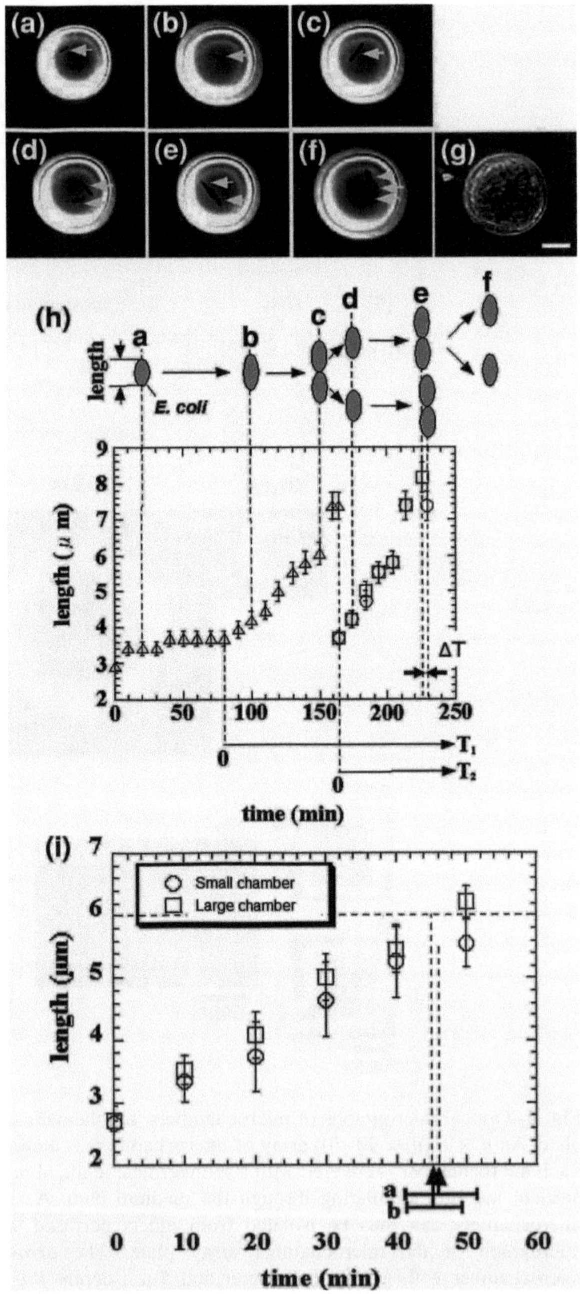

Fig. 5 Differential analysis of sister cells from isolated single *E. coli* in a microchamber. The magnified micrographs at the top **a–g** show the time course of one of the microchambers in Fig. 4b (see *arrow*) at times of **a** 0 min, **b** 100 min, **c** 150 min, **d** 175 min, **e** 225 min, **f** 230 min, and **g** 15 h after the inoculation started. The *arrows* in the micrographs show the positions of *E. coli* in the microchamber. *Bar* 10 μm. **h** Time course growth of the individual *E. coli*. **i** Chamber size dependence of cell growth and interdivision time, small chamber and large chamber in Fig. 4c (*open circle* and *open square* for cell length, and *a* and *b* for interdivision time, respectively), indicating no dependence of cell growth speed on chamber size

under isolated conditions. Four specific cells in the four compartments were simultaneously observed without any disruption by the other cells and without leaving the field of view of the microscope.

Optical tweezers were introduced to enable non-contact handling of the cell specimens. An Nd:YAG laser (wavelength = 1,064 nm, T20-8S, Spectra Physics, SpectraPhysics, CA) was guided to the ×100 phase contrast objective lens (UplanApo, Olympus, Tokyo, Japan) as the light source for the optical tweezers, which are widely used in handling micron-sized particles and biomaterials [44–49]. We used it in the system in our protocol to transport particular cells within the microchamber.

The medium circulation unit utilized a glass box with a volume of 1 ml that had two branches. It was mounted on the microchamber array chip and fresh medium buffer was always circulated in the glass box through the two branches at a rate of 1 ml/min with a peristaltic pump. The bottom of the glass box was open and the condition of the medium around the cells could be constantly maintained by buffer exchanges through the semipermeable membrane.

A phase contrast microscope (obj. ×100 magnitude) was set up with IX-70 (Olympus). The whole microcultivation part was placed in a thermo control cage (IX-IBM, Olympus) to maintain the temperature at 37 °C throughout observation. The observation images were taken with a CCD camera (CS230, Olympus) and recorded on digital video cassette. These were analyzed on a personal computer (PCV-R73 K, Sony, Tokyo, Japan).

2.2 *Differential Analysis of Sister Cells with Identical Genetic Information and Experience*

To investigate non-genetic variability in the division cycle and growth of single cells, we first compared the growth and division times for pairs of *E. coli* daughter cells under isolated conditions using the on-chip single-cell cultivation system we just have described [3, 5] (Figs. 3 and 4).

In this experiment, we used *E. coli* strain JM109 (*end*A1, *rec*A1, *gyr*A96, *thi*, *hsd*R17(r_k^-, m_k^+), *rel*A1, *sup*E44, λ^-, Δ(*lac-pro*AB), [F′, *tra*D36, *pro*AB, *lac*I^{q}ZΔM15] obtained from Toyobo, Tokyo, Japan) in a minimal medium, M9 (4.5 g/l KH_2PO_4, 10.5 g/l K_2HPO_4, 50 mg/l $MgSO_4{\cdot}7H_2O$, pH 7.1) containing 0.2% (w/w) of glucose at 37 °C.

First, we checked the chamber size dependence of cell growth and interdivision time using different sized microchambers (Fig. 4c) and found there no apparent difference (Fig. 5i).

Before starting continuous isolation of direct descendant cells using optical tweezers, we have compared the difference of growth and divisions of sister cells within the microchambers under the conditions without any physical stimulation applied to the cultivating cells. After on-chip single-cell cultivation has started, an isolated single cell (mother cell) grew in the microchamber after the resting of growth from 2.8 to 5.6 μm in 90 min, and finally divided into two 2.8-μm daughter cells (Fig. 6a). Although the newborn daughter cells grew synchronously in the same manner, they divided into granddaughter cells at different times, i.e., 70 and

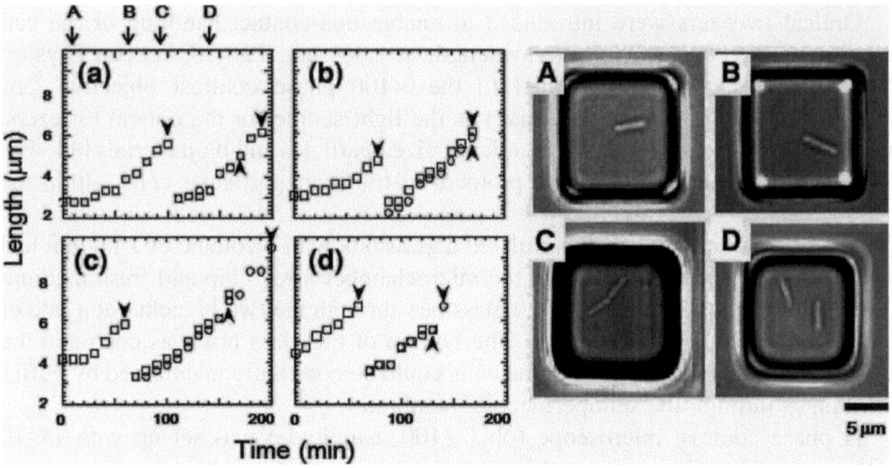

Fig. 6 Time course growth for isolated individual *E. coli* and two daughters

90 min (see arrowheads in graph). The three other examples (Fig. 6b–d) show that even though the growth of the mother cell and her daughter cells seems to have no significant correlation, the growth of two daughter cells from the same mother cell seems to be quite similar. In contrast, the division times for daughter cells of the same length (Fig. 6a, c, d) were not synchronous. In Fig. 6b, on the other hand, the division time and cell growth tendency of two daughter cells were synchronous even though they were born after unequal divisions of the mother cell. These results indicate that variations in cell growth and cell division may not be closely correlated and that cell division time is independent of genetic identity and cell size.

The division time differences between two daughter cells from the same mother cells were also measured (Fig. 7). Although sister cells are thought to have the same DNA and chemical components as their mother's cells, the results revealed only 36% of daughter cells divided into granddaughter cells within a 10-min difference of period even when they started at the same cell lengths (Fig. 7a). The dependence of division time differences for newborn daughter cells on length was also evaluated and the time distribution was similar regardless of the initial length (Fig. 7b). These results mean that variations in cell division may not depend on DNA or the initial cell size.

The initial dependence of variations in cell growth and division on length was also evaluated. The ratio of the final length of these cells and their initial length seems to be independent of the initial length, about 170%, when it is longer than 3 μm. The speed of growth of cells also has no significant dependence on the initial length.

In this experiment, we observed and compared the cell growth and division of two daughter cells of isolated single *E. coli* using the on-chip culture system, and found a broad range of variations in cell growth and division time. Such variations are not attributable to the genetic differences in DNA. The variations in the growth

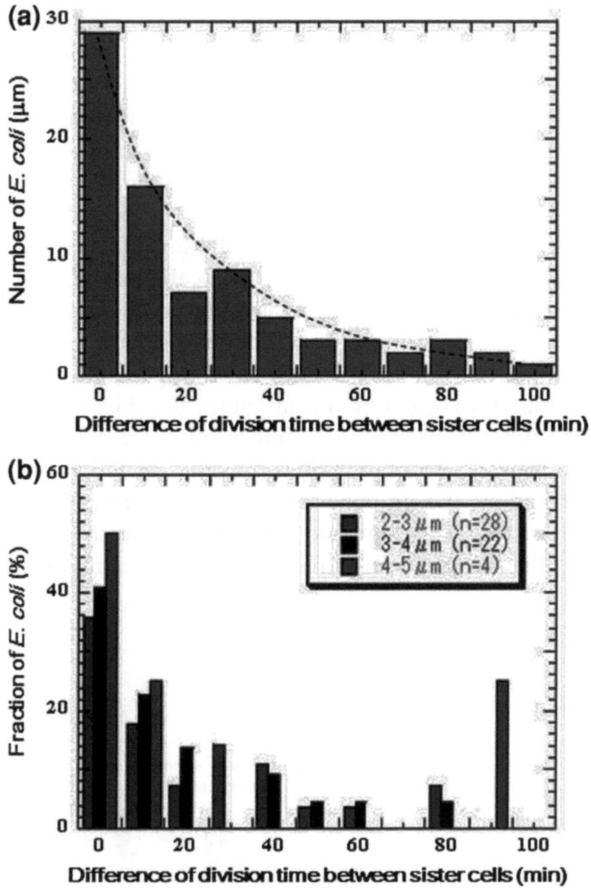

Fig. 7 **a** Differences in division time for two daughter cells of same mother cells ($n = 80$ pairs), and **b** initial dependence of division time differences on length

ratio between the final and initial lengths, and the speed of growth seem to be independent of their initial length, at least when they are longer than 3 μm. The same tendency toward a broad distribution in the division time (data not shown) and the division time differences of two daughter cells from the same mother cells (Fig. 7) suggests the involvement of a probabilistic process that starts division. A Poissonian variation in a small number of molecules that determines growth and division might explain the origin of these non-genetic variations in cells.

2.3 *Differential Analysis of Direct Descendant Cells with Identical Genetic Information and Experience*

We next examined whether the characteristics of direct descendants of an isolated single cell could be inherited under isolated conditions using the on-chip single-cell cultivation/analysis system with optical trapping to maintain the isolated condition of cells even after several generations of cell divisions [4, 6]. Figure 8 plots temporal variations in cell lengths of individuals and their descendants. Figure 8a shows schematics explaining the measured interdivision time and cell length. The four graphs (b–e) indicate growth and division patterns for four cells born from a single cell and isolated into the four chambers A–D in Fig. 4f.

Figure 9a also plots variations in interdivision times for consecutive generations of other isolated *E. coli* cells derived from a common ancestor. The four series of interdivision times varied around the overall mean value, 52 min (dashed line); the mean values of the four cell lines a, b, c, and d were 54, 51, 56 and 56 min, indicating rather small differences compared with the large variations in the interdivision times of consecutive generations. These results support the idea that interdivision time variations from generation to generation are dominated by fluctuations around the mean value, and this was evidence of a stabilized phenotype that was subsequently inherited. To explore this idea further, we examined the dependence of interdivision time on the interdivision time of the previous generation. We grouped all interdivision time data into four categories and calculated their distributions (Fig. 9b). A comparison of these distributions revealed that they were astonishingly similar, suggesting that there was no dependence on

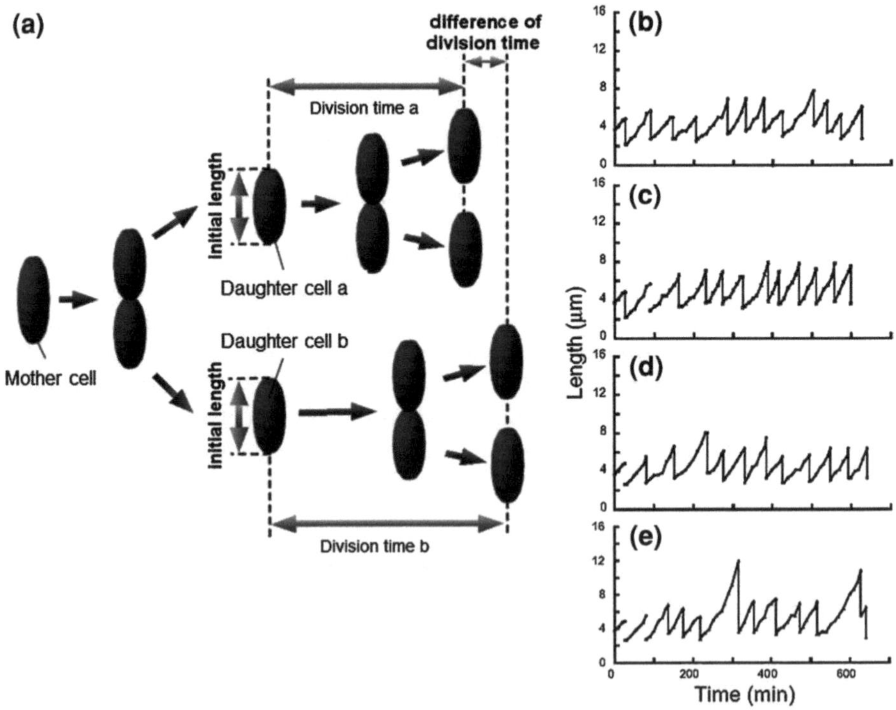

Fig. 8 Temporal variations in cell lengths of individual cells and their direct descendants

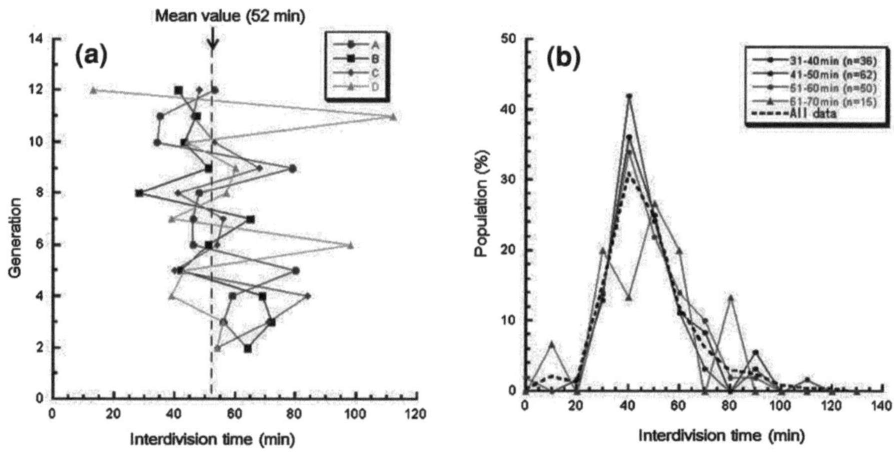

Mean values for 10 generations in each chamber

Chamber	A	B	C	D
Mean (min)	54	51	56	56

within (total mean time) ±4 min

Fig. 9 Variations of interdivision time in direct descendant cells of *E. coli*

the previous generation. That is, there was no inheritance of interdivision time from one generation to the next.

2.4 Adaptation Process for Sensor Proteins in Cells Caused by Environmental Changes

We have modified the on-chip single-cell cultivation/analysis system to simultaneously measure the sensor-protein dynamics and motility of identical single cells for several generations [8]. This technique revealed the potential of combining the microfabrication technique (single-cell cultivation technique) and molecular biology (single-molecule observation).

Escherichiacoli cells are able to respond to changes in environmental chemoeffector concentrations through reversing their flagellar motors [50, 51]. Attractants (such as aspartate and serine) promote counterclockwise rotation of the flagella, resulting in a smooth swimming action, whereas repellents (such as phenol and Ni) promote clockwise rotation, resulting in tumbling. These responses are mediated by membrane-bound, methyl-accepting chemoreceptor proteins (MCPs). Immunoelectron microscopy revealed that MCP–CheW–CheA complexes are clustered in vivo, predominantly at the cell poles [52], and merely weaker lateral clusters could be observed [53–55]. Polar localization changes have been expected according to environmental conditions, whereas no evidence concerning the dynamics of localization changes has been reported. Conventional group-based experiments do not allow the process of MCP clustering and the effect its change has on consecutive generations in individual cells, which is essential in estimating the changes occurring during the alternation of generations. To understand epigenetic processes such as adaptation and selection, both the protein dynamics and the cell dynamics of particular single cells should be observed continuously and simultaneously for several generations.

We used assayed intracellular proteins tagged with green fluorescent protein (GFP) to measure the localization dynamics of expressed proteins (Fig. 10).

We modified the shape of the microchambers into a wheel to measure the time course for motility (Fig. 11a). In the experiment, we first placed a single bacterium in the microchamber and isolated it in the wheel region so that it could swim along the track seal with the semipermeable membrane lid on the microchamber. Then, the bacterium running around the circle structure was continuously monitored by measuring the tumbling frequency and protein localization dynamics. When the cell divided into two daughter cells, one of these was picked up with the optical tweezers, transported to the axle area, and continuously confined in this region to stop it growing. The bacterium was chemically stimulated by changing the contents of the medium.

When the cultivation started, the tar localization ratio (filled squares) was 2.5 and the tumbling frequency (filled circles) was 0.5 (s^{-1}) (Fig. 11B-a and arrowhead 'a' in Fig. 11C). After the second cell division had occurred, a minimal medium containing 1 mM of aspartate was applied to the third-generation cell (135 min after microcultivation). After the attractant was added, tumbling frequency decreased immediately compared to the previous generation. Localization of the aspartate-sensitive sensor protein at two poles in *E. coli* (filled squares) also decreased quickly by half to 45 min following the change of medium (Fig. 11B-b, C-b). Finally, after 80 min of stimulation with the aspartate,

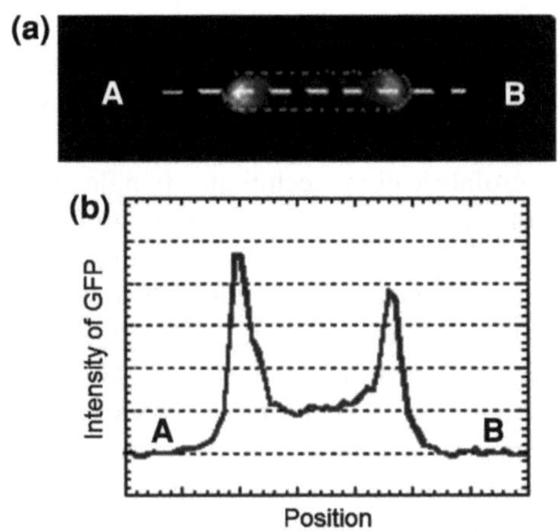

Fig. 10 MCP–GFP localization in *E. coli* cell. **a** Fluorescent micrograph of *E. coli*. **b** Intensity profile of MCP–GFP localization

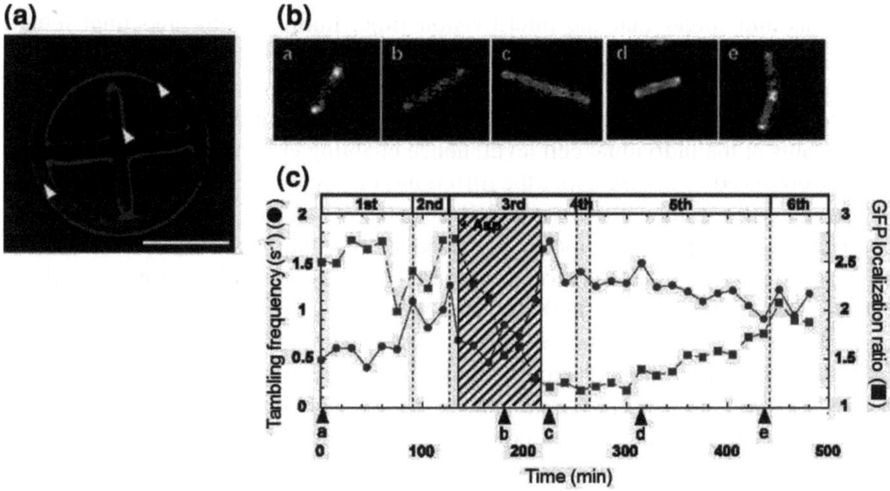

Fig. 11 Simultaneous observation of MCP–GFP localization and motility in identical *E. coli* cell for generations

the localized tar had diffused completely. Then, the aspartate was removed from the cultivation medium and the cells were cultivated further to enable the recovery of tar localization dynamics to be measured (Fig. 11B-c, C-c). After the first change of medium, it took more than three generations to recover the original pattern of tar localization (Fig. 11B-d, B-e, C-d, C-e). However, the frequency of tumbling remained higher than the former generations. This may indicate that tar localization requires more time to form than to diffuse. Such asymmetric reversibility in protein localization may contribute to cell phenomena being inherited in response to environmental changes [1, 4, 7, 56–66]. It also suggests the possibility that change in tar localization can be inherited by descendant cells and this can affect their motility and therefore their phenotype.

2.5 Epigenetic Inheritance of Elongated Phenotypes Between Generations Revealed by Individual-Cell-Based Direct Observation [67–76]

When implementing individual-cell-based measurements of epigenetic inheritance instead of group-based measurement, we developed an 'on-chip individual-cell cultivation system' [3, 4, 77]. This system enabled us to compare the phenotypes between generations and to examine the existence of phenotypic transmission at the individual cell level under stringently controlled conditions. We called this method a 'differential individual-cell observation assay'. We measured the variations in quantitative traits at the individual cell level under uniform and isolated conditions and found that the interdivision time, initial length, and final length varied at the individual cell level as much as 33, 26, and 26%, respectively [78].

This experiment reports the phenotypic dynamics involved in changing normal isolated cells into elongated ones, and the transmission of elongated phenotypes to descendants at the individual cell level, hence enabling epigenetic inheritance to be directly observed in *E. coli* using the differential individual-cell observation assay.

The *E. coli* strain EJ2848 (LacI3 ΔlacZ lacY+ΔfliC) [79] was used in this experiment. To simplify the model, we used a strain that lacks motility due to the deletion of flagellin. Cells were prepared by using M9 minimal medium (Qbiogene) supplemented with 0.2% (w/w) glucose and amino acids (MEM amino acids, Invitrogen). The cultivated cells were loaded and observed in the on-chip individual-cell cultivation system described in a previous report [78].

Figure 12a plots example growth and division dynamics of individual *E. coli* cells under uniform and isolated conditions. A single cell was first loaded in the on-chip individual-cell cultivation system and observed continuously (Fig. 12a). The cell exhibited normal growth and division patterns in the first and second

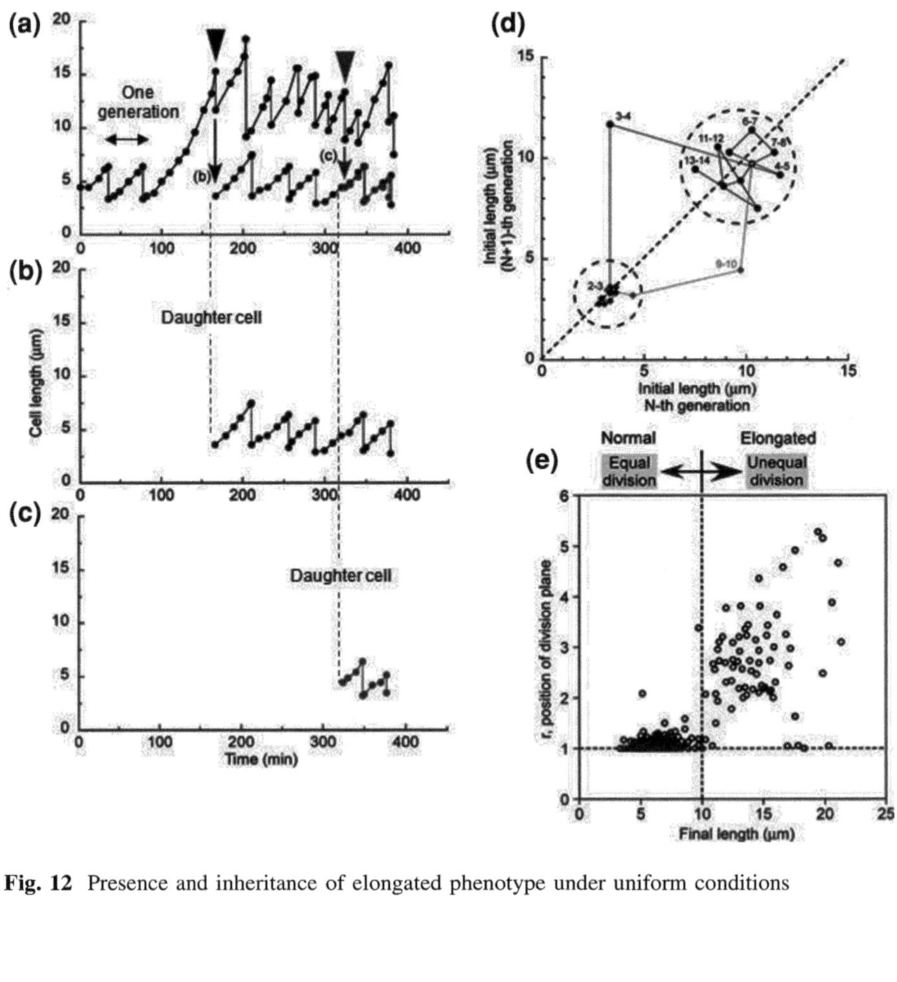

Fig. 12 Presence and inheritance of elongated phenotype under uniform conditions

generations and divided almost equally; hence, we randomly discarded one of the two daughter cells (closest to the exit). Despite the normal growth and division characteristics in the first and second generations, the cell in the third generation was extraordinarily elongated. A striking feature of the cell was that it divided unequally (Fig. 12a, arrowheads), thereby producing two daughter cells; one was elongated and the other was normal (Fig. 12a). Elongated daughter cells followed in the subsequent descendents, remaining elongated through repeated unequal cell divisions in most cell divisions. In other words, the cell transmitted its abnormally long cell length to its descendants. However, the normal daughter exhibited normal growth and division patterns and did not elongate in the following generations (Fig. 12b). The normal daughter cell born from the elongated cell had the same tendencies as those of typical normal cells. Just as in the third generation, the shorter of the two daughter cells in the ninth generation also demonstrated normal growth and division patterns (Fig. 12c). Therefore, it is conceivable that the shorter of the two daughters from the elongated cells possesses a normal phenotype.

The transmission of the elongated phenotype between generations can clearly be seen in the returning map of initial cell-length transitions (Fig. 12d). The converging dots indicate stable states in the transition. The initial cell-length transition for elongated lineage jumped from the normal cell-length area to the elongated cell-length area at the beginning of the fourth generation and stayed there throughout the following generations, whereas the shorter daughters' transitions were within the normal cell-length area. The clear distinction between the two transition areas hence indicates that the elongated cell transmitted a distinctive phenotype to one lineage of its descendants.

The presence of elongated cells was not restricted to this example. We found 5% of the normal-phenotype cells (12 out of 242) observed in the on-chip individual-cell cultivation system under the same conditions acquired elongated phenotypes.

The question is what is the mechanism responsible for passing on the elongated phenotype in one lineage of descendants? It should be noted that the evidence accrued from our results suggests the presence of elongated cells is not caused by the mutation of genetic information because one of the two daughters possessed normal cell characteristics, which should have had the same genetic information as the other elongated daughter cells. Otherwise, the normal of the two daughters would have elongated in the following generations if the elongation had been caused by mutation.

The repeated unequal divisions of elongated cells in Fig. 12a suggest that there is an intracellular mechanism that induces unequal cell divisions to elongate cells. Moreover, the mechanism for inducing unequal cell division should underlie the stable inheritance of the elongated phenotype in one lineage; unequal cell division produces a daughter cell with a long start cell, which would also attain a long cell by the next division, enabling it to divide unequally again. Repeating this process, a cell could stably transmit the elongated phenotype to the one lineage of descendants once they acquired a long cell.

The question arising from the above observations is whether there is a boundary for length that changes cell characteristics if variations exceed a certain length. We then examined the relationship between the final length and the position of the division plane (Fig. 12e). The position of the division plane, r, was defined by using the cell lengths of two daughter cells produced in the corresponding cell divisions as

$$r = \frac{\text{(Initial length of longer daughter cell)}}{\text{(Initial length of shorter daughter cell)}}. \quad (1)$$

Therefore, '$r = 1$' corresponds to equal cell division and '$r > 1$' corresponds to unequal cell division. Figure 12e shows that the position of the division plane was at an uneven point when the cell length was longer than 10 μm. The results clearly reveal that there is a boundary for cell length that affects division characteristics; cells shorter than 10 μm divide equally (normal phenotypes), while those longer than 10 μm divide unequally (elongated phenotypes). Hence, it is conceivable that a kind of geometric index, i.e., cell length, controls their phenotypic characteristics.

We next examined the division cycles of elongated cells. Figure 13a plots the interdivision time distributions for elongated and normal cells. 'Elongated cells' were defined as those whose final length was longer than 10 μm and those whose final length in the previous generation was also longer than 10 μm. Based on this definition, the interdivision time for generation in which a cell elongated extraordinarily from normal length (like the third generation in Fig. 12a) was not categorized as elongated in plotting the distribution. The average elongated-cell interdivision time was 25.9 ± 1.6 min, which was half that of normal cells (52.4 ± 1.1 min) and might be able to explain how the cell division process of both ends of the elongated cell proceeded independently. The coefficients of variance (CV) were 48 and 33% for the elongated and the normal, respectively. The distinct interdivision time distribution for elongated cells also confirms that they could easily be distinguished from normal cells. The characteristics of elongated cells cannot be explained by variations in normal cells. We thus regarded elongated and normal cells to be different phenotypes.

The next question is how does an elongated cell achieve a short interdivision time? Figure 13b shows an alternative formation for the division plane at opposite poles between neighboring generations. Image 1 shows an elongated cell emerging in the fourth generation from a normal phenotype before the division plane is formed. A visible division plane was then formed near the lower end as can be seen from Image 2 (fourth generation). After cell division, the longest daughter cell was selected and observed thereafter. We found a division plane was formed in the next generation near the opposite end of the cell (Image 3, fifth generation). Although the cell divided equally and produced two elongated daughters in the sixth generation, the division ended in the following two generations (seventh and eighth) exhibiting the same behavior as in the fourth and fifth generations, i.e., division planes were generated on opposite sides (Images 4 and 5).

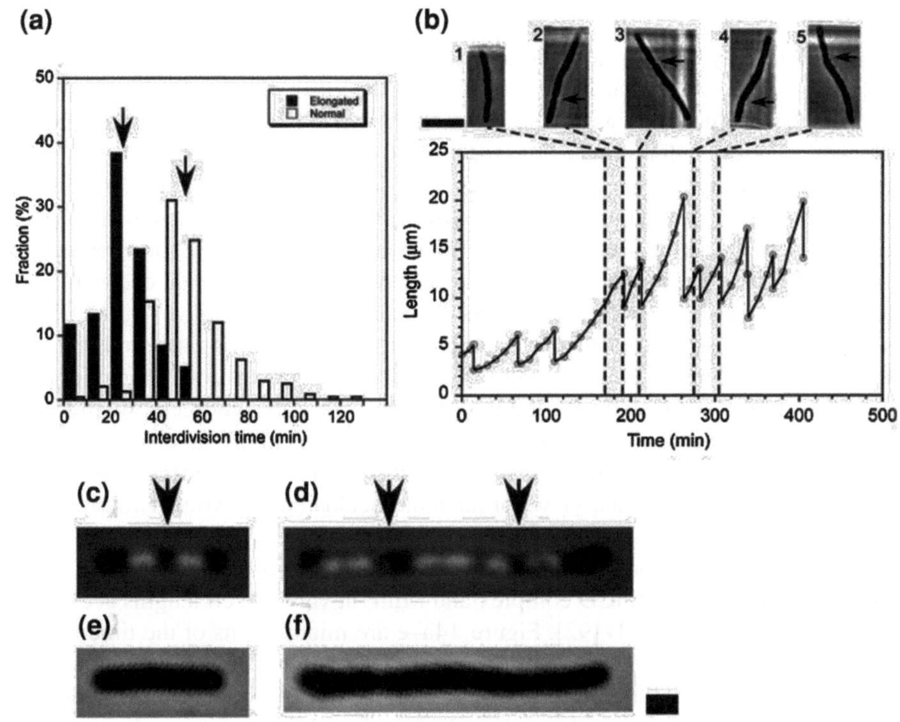

Fig. 13 Short interdivision time for elongated cell. *Bar* 1 μm

The fact that the positions of division planes in consecutive generations (fourth–fifth and seventh–eighth) were opposite indicates that two division mechanisms operate simultaneously in an elongated cell near both ends. The half interdivision time of elongated cells can be explained by two division planes simultaneously and independently working near both ends in elongated cells, each of which works at approximately the same interval as interdivision in the normal phenotype.

The question then is what intracellular mechanism is responsible for the observed unequal cell divisions in elongated cells? It has been suggested that two mechanisms in *E. coli*, nucleoid occlusion and the Min system, determine the position of the division plane. Nucleoid occlusion is the mechanism where the assembly of the FtsZ ring, which determines the position of the final cell division plane, is inhibited within the close vicinities of nucleoids [80, 81]. The Min system, on the other hand, is comprised of three proteins, MinC, MinD, and MinE, which dynamically interact with one another and exhibit rapid pole-to-pole oscillations [82–92]. MinC and MinD form an inhibitory complex to form Z rings [85–87]; Z-ring formation has been proposed to be directed to positions where concentrations of the MinCD complex are low on average in oscillations [89, 91, 92]. To examine the relevance of observed unequal cell divisions to the proposed mechanisms responsible for the position of the division plane, we visualized the position of nucleoids in elongated cells by 4′,6-diamidino-2-phenylindole dihydrochloride hydrate (DAPI) fluorescence using Hiraga's method [93] (Fig. 13c, d). A normal-sized cell reaching the mean division length possessed two segregated nucleoids with a space in the middle as seen in Fig. 13c (phase-contrast image for reference in Fig. 13e). MinCD concentration should be low in the middle and high at the poles [89, 91, 92], leading to Z-ring formation in the middle. However, an elongated cell possessed multiple nucleoids with spaces between them as can be seen from Fig. 13d (phase-contrast image for reference in Fig. 13f). There are numerous positions where constrictions can occur in this elongated cell according to the nucleoid occlusion model. However, MinCDE oscillations are known to achieve 'doubled' patterns in an elongated cell [85, 86, 88], which has been proposed to inhibit Z-ring formation both around the midplane and at the poles. Consequently, the cell divisions in elongated cells of this size should only occur at the uneven nucleoid gap positions (Fig. 13d, arrows).

Although the presence of unequal cell divisions in elongated cells is understandable from the combined views of nucleoid occlusion and Min oscillation, it is still unclear what determines the cell-length boundary, 10 μm, between the equal and unequal cell divisions shown in Fig. 12e. We therefore calculated the oscillation dynamics of the MinCD complex and MinE at various cell lengths according to Huang's model (Fig. 14) [92]. Figure 14a–e are micrographs of the time course change in the concentration of MinD at the cell membrane along the long axis of the cell, indicating that the number of oscillations doubles when the cell length is longer than 8 μm. The single-cycle averages for MinD concentration at the membrane in Fig. 14a–e also reveal the concentration of MinD at the midplane becomes highest for all positions when the cell length is longer than 8 μm. The MinD concentration at the midplane relative to the overall averages is plotted in Fig. 14f. The results indicate a drastic increase in MinD concentration at the midplane when the oscillation doubles, which suggests that the cell-length boundary between the single and double oscillations of Min proteins determines the cell-length boundary between equal and unequal cell divisions in Fig. 12e.

On the basis of the results in Fig. 14f, FtsZ should be unable to form a ring at the midplane in cells with lengths longer than 8 μm. Therefore, FtsZ would inevitably form rings at uneven positions in these long cells, which would lead to unequal cell divisions. The formation of FtsZ rings lies at the heart of the process of division at the membrane [94]. In previous studies, the time between when cell division was initiated by the polymerization of FtsZ and the appearance of constriction visualized by electron microscopy was found to be approximately 20 min [95]. Under the conditions in our experiment, the average time to double cell length was 52 min (Fig. 13a); hence, the expected division cell-length for a cell whose Z-ring formation was initiated at a cell length of 8 μm can be calculated as $8\ \mu m \times 2^{20/52} = 10.4\ \mu m$. This calculation suggests that the boundary for final length that separates equal and unequal cell divisions should be 10.4 μm, which matches our experimental results in Fig. 12e. All these results indicate that it is highly probable that Min oscillations determine the observed cell-length boundary between equal and unequal cell divisions.

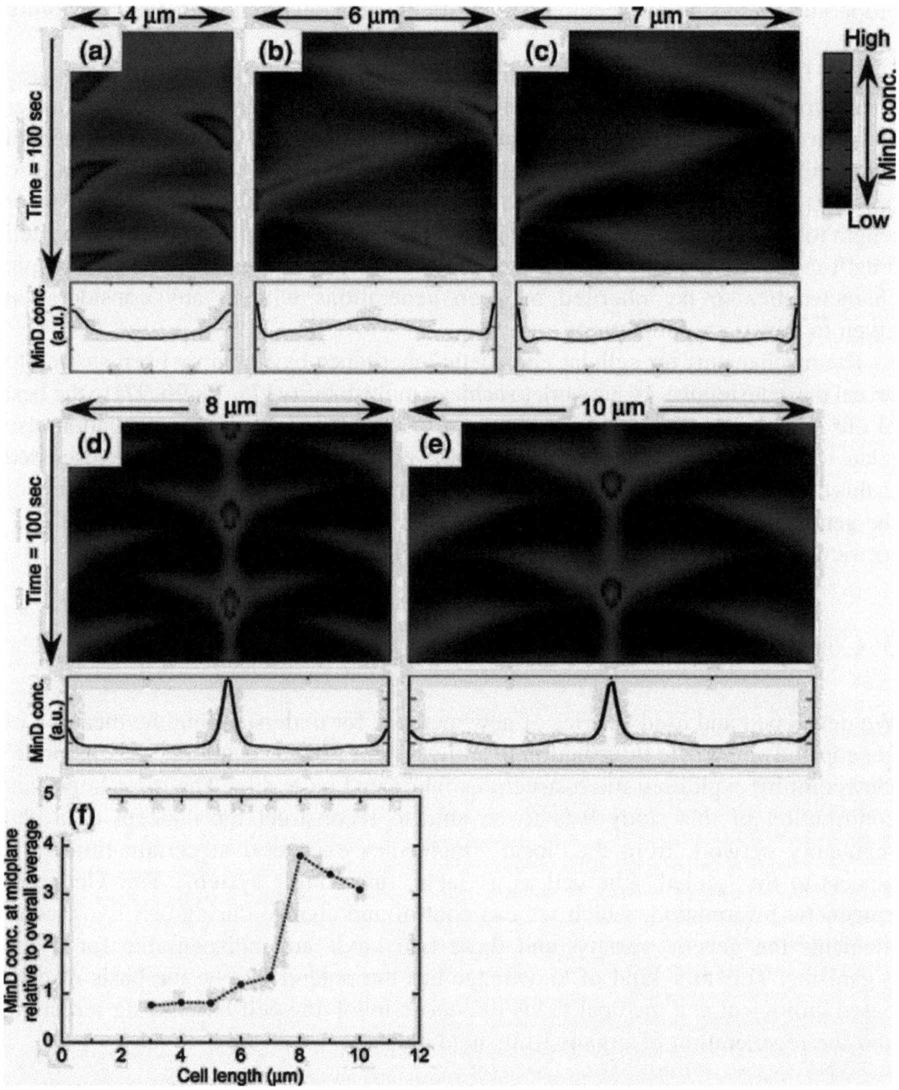

Fig. 14 Simulated MinD dynamics within cells with various cell lengths

It is still unclear at this point how a cell acquires length. We previously reported that the length distributions for the division of genetically identical cells in the same environment had variations of 26% (CV) [78]. The large variations in cell-length distributions reflect uncertainty about whether genetically identical cells can generate the same length when the environmental conditions are the same. In other words, the intracellular mechanism inevitably includes stochasticity, which causes variations in the length of divided cells. Therefore, extraordinarily elongated cells with final lengths longer than 10 µm can occur with a certain probability based on intracellular stochasticity; this occurred with 5% probability under our test conditions.

The fact that elongated cells divide unequally would enable the elongated phenotype to be inherited in one lineage. Unequal cell division produces longer and shorter daughter cells. The longer daughter cell with a long starting cell length can easily exceed the boundary cell length. Consequently, the next division also becomes unequal. Repeating this process, a cell will eventually transmits its longer length to one of its descendant once it has acquired that length. Geometry, i.e., cell length in this inheritance mechanism, plays a key role in enabling the phenotypic characteristics to be inherited between generations without any consideration given to genetic modifications.

The mechanisms for cellular epigenetic inheritance have mainly been studied to reveal the gene regulatory network to achieve multistability [75, 76, 96, 97] to the best of our knowledge. The roles of chromatin, DNA methylation, and acetylation also relate to gene regulation [74, 98–100]. However, the epigenetic inheritance discussed in this chapter occurred once a cell reached a certain cell length; it was independent of the gene regulatory network. Therefore, our results suggest the inheritance of geometric information, such as cell shape, is significant in epigenetic inheritance.

3 Conclusion

We developed and used a series of new methods for understanding the meaning of genetic and epigenetic information in a life system from a 'temporal' or 'algebraic' viewpoint by exploiting microstructures fabricated on a chip. The most important contribution of this study was to be able to reconstruct the concept of a cell regulatory network from the 'local' (molecules expressed at certain times and places) to the 'global' (the cell as a viable, functioning system). Knowledge of epigenetic information, which we can control and change during cell lives, complements the genetic variety, and these two kinds are indispensable for living organisms. This new kind of knowledge has the potential to be the basis of cell-based biological and medical fields like those involving cell-based drug screening and the regeneration of organs from stem cells.

Acknowledgments The author acknowledges the assistance of all members of Yasuda Lab. Financial support, in part from the Japan Science and Technology Agency (JST) and from Grants-in-Aids for Scientific Research from the Ministry of Education, Culture, Sports, Science and Technology of the Japanese government, is gratefully appreciated.

References

1. Spudich JL, Koshland DE Jr (1976) Non-genetic individuality: chance in the single cell. Nature 262:467–471
2. Yasuda K (2000) Non-destructive, non-contact handling method for biomaterials in micro-chamber by ultrasound. Sens Actuators B Chem 64:128–135

3. Inoue I, Wakamoto Y, Moriguchi H, Okano K, Yasuda K (2001) On-chip culture system for observation of isolated individual cells. Lab Chip 1:50–55
4. Wakamoto Y, Inoue I, Moriguchi H, Yasuda K (2001) Analysis of single-cell differences using on-chip microculture system and optical trapping. Fresenius J Anal Chem 371:276–281
5. Inoue I, Wakamoto Y, Yasuda K (2001) Non-genetic variability of division cycle and growth of isolated individual cells in on-chip culture system. Proc Jpn Acad 77B:145–150
6. Wakamoto Y, Umehara S, Matsumura K, Inoue I, Yasuda K (2003) Development of non-destructive, non-contact single-cell based differential cell assay using on-chip microcultivation and optical tweezers. Sens Actuators B Chem 96:693–700
7. Umehara S, Wakamoto Y, Inoue I, Yasuda K (2003) On-chip single-cell microcultivation assay for monitoring environmental effects on isolated cells. Biochem Biophys Res Commun 305:534–540
8. Inoue I, Shiomi D, Kawagishi I, Yasuda K (2004) Simultaneous measurement of sensor-protein dynamics and motility of a single cell by on-chip microcultivation system. J Nanobiotechnology 2:4
9. Takahashi K, Matsumura K, Yasuda K (2003) On-chip microcultivation chamber for swimming cells using visualized poly(dimetylsiloxane) valve. Jpn J Appl Phys 42:L1104–L1107
10. Hattori A, Umehara S, Wakamoto Y, Yasuda K (2003) Measurement of incident angle dependence of swimming bacterium reflection using on-chip single-cell cultivation assay. Jpn J Appl Phys 42:L873–L875
11. Matsumura K, Yagi T, Yasuda K (2003) Role of timer and sizer in regulation of *Chlamydomonas* cell cycle. Biochem Biophys Res Commun 306:1042–1049
12. Matsumura K, Yagi T, Yasuda K (2003) Differential analysis of cell cycle stability in *Chlamydomonas* using on-chip single-cell cultivation system. Jpn J Appl Phys 42:L784–L787
13. Moriguchi H, Wakamoto Y, Sugio Y, Takahashi K, Inoue I, Yasuda Y (2002) An agar-microchamber cell-cultivation system: flexible change of microchamber shapes during cultivation by photo-thermal etching. Lab Chip 2:125–30
14. Hattori A, Moriguchi H, Ishiwata S, Yasuda K (2004) A 1480-nm/1064-nm dual wavelength photo-thermal etching system for non-contact three-dimensional microstructure generation into agar microculture chip. Sens Actuators B Chem 100:455–462
15. Sugio Y, Kojima K, Moriguchi H, Takahashi K, Kaneko T, Yasuda K (2004) An agar-based on-chip neural-cell cultivation system for stepwise control of network pattern generation during cultivation. Sens Actuators B Chem 99:156–162
16. Moriguchi H, Takahashi K, Sugio Y, Wakamoto Y, Inoue I, Jimbo Y, Yasuda K (2003) On chip neural cell cultivation using agarose-microchamber array constructed by photo-thermal etching method. Electr Eng Jpn 146:37–42
17. Suzuki I, Sugio Y, Jimbo Y, Yasuda K (2004) Individual-cell-based electrophysiological measurement of a topographically controlled neuronal network pattern using agarose architecture with a multi-electrode array. Jpn J Appl Phys 43(3B):L403–L406
18. Suzuki I, Sugio Y, Jimbo Y, Yasuda K (2004) Modification of a neuronal network direction using stepwise photo-thermal etching of an agarose architecture. J Nanobiotechnology 2:7
19. Kojima K, Moriguchi H, Hattori A, Kaneko T, Yasuda K (2003) Two-dimensional network formation of cardiac myocytes in agar microculture chip with 1480-nm infrared laser photo-thermal etching. Lab Chip 3:299–303
20. Kojima K, Kaneko T, Yasuda K (2006) Role of the community effect of cardiomyocyte in the entrainment and reestablishment of stable beating rhythms. Biochem Biophys Res Commun 351:209–215
21. Yasuda K, Okano K, Ishiwata S (2000) Focal extraction of surface-bound DNA from a microchip using photo-thermal denaturation. Biotechniques 28:1006–1011
22. Gally DL, Bogan JA, Eisenstein BI, Blomfield IC (1993) Environmental regulation of the fim switch controlling type 1 fimbrial phase variation in *Escherichia coli* K-12: effects of temperature and media. J Bacteriol 175:6186–6193

23. Ko MS, Nakauchi H, Takahashi N (1990) The dose dependence of glucocorticoid-inducible gene expression results from changes in the number of transcriptionally active templates. EMBO J 9:2835–2842
24. Msadek T (1999) When the going gets tough: survival strategies and environmental signaling networks in *Bacillus subtilis*. Trends Microbiol 7:201–207
25. Schwan WR, Seifert HS, Duncan JL (1992) Growth conditions mediate differential transcription of fim genes involved in phase variation of type 1 pili. J Bacteriol 174: 2367–2375
26. Shapiro JA (1995) The significances of bacterial colony patterns. Bioessays 17:597–607
27. Shapiro JA (1998) Thinking about bacterial populations as multicellular organisms. Annu Rev Microbiol 52:81–104
28. Shapiro JA, Hsu C (1989) *Escherichia coli* K-12 cell–cell interactions seen by time-lapse video. J Bacteriol 171:5963–5974
29. Arkin A, Ross J, McAdams HH (1998) Stochastic kinetic analysis of developmental pathway bifurcation in phage lambda-infected *Escherichia coli* cells. Genetics 149:1633–1648
30. Kepler TB, Elston TC (2001) Stochasticity in transcriptional regulation: origins, consequences, and mathematical representations. Biophys J 81:3116–3136
31. Kierzek AM, Zaim J, Zielenkiewicz P (2001) The effect of transcription and translation initiation frequencies on the stochastic fluctuations in prokaryotic gene expression. J Biol Chem 276:8165–8172
32. Levin MD, Morton-Firth CJ, Abouhamad WN, Bourret RB, Bray D (1998) Origins of individual swimming behavior in bacteria. Biophys J 74:175–181
33. McAdams HH, Arkin A (1997) Stochastic mechanisms in gene expression. Proc Natl Acad Sci USA 94:814–819
34. Thattai M, van Oudenaarden A (2001) Intrinsic noise in gene regulatory networks. Proc Natl Acad Sci USA 98:8614–8619
35. Blake WJ, Kaern M, Cantor CR, Collins JJ (2003) Noise in eukaryotic gene expression. Nature 422:633–637
36. Elowitz MB, Levine AJ, Siggia ED, Swain PS (2002) Stochastic gene expression in a single cell. Science 297:1183–1186
37. Ozbudak EM, Thattai M, Kurtser I, Grossman AD, van Oudenaarden A (2002) Regulation of noise in the expression of a single gene. Nat Genet 31:69–73
38. Rao CV, Wolf DM, Arkin AP (2002) Control, exploitation and tolerance of intracellular noise. Nature 420:231–237
39. Åkerlund T, Nordström K, Bernander R (1995) Analysis of cell size and DNA content in exponentially growing and stationary-phase batch cultures of *Escherichia coli*. J Bacteriol 177:6791–6797
40. Donachie DD, Begg KJ (1996) 'Division potential' in *Escherichia coli*. J Bacteriol 178:5971–5976
41. Elowitz MB, Leibler S (2000) A synthetic oscillatory network of transcriptional regulators. Science 403:335–338
42. Gardner TS, Cantor CR, Collins JJ (2000) Construction of a genetic toggle switch in *Escherichia coli*. Science 403:339–342
43. Panda AK, Khan RH, Appa Rao KBC, Totey SM(1999) Kinetics of inclusion body production in batch and high cell density fed-batch culture of *Escherichia coli* expressing ovine growth hormone. J Biotech 75:161–172
44. Ashkin A, Dziedzic JM, Bjorkholm JE, Chu S (1986) Observation of a single-beam gradient force optical trap for dielectric particles. Opt Lett 11:288–290
45. Ashkin A, Dziedzic JM, Yamane T (1987) Optical trapping and manipulation of single cells using infrared laser beams. Nature 330:769–771
46. Ashkin A, Dziedzic JM (1989) Internal cell manipulation using infrared laser traps. Proc Natl Acad Sci USA 86:7914–7918

47. Ashkin A, Schutze K, Dziedzic JM, Euteneuer U, Schliwa M (1990) Force generation of organelle transport measured in vivo by an infrared laser trap. Nature 348:346–348
48. Wright W, Sonek GJ, Tadir Y, Berns MW (1990) Laser trapping in cell biology. IEEE J Quantum Electron 26:2148–2157
49. Svoboda K, Block SM (1994) Biological applications of optical forces. Annu Rev Biophys Biomol Struct 23:247–285
50. Levit MN, Liu Y, Stock JB (1998) Stimulus response coupling in bacterial chemotaxis: receptor dimers in signalling arrays. Mol Microbiol 30:459–466
51. Manson MD, Armitage JP, Hoch JA, Macnab RM (1998) Bacterial locomotion and signal transduction. J Bacteriol 180:1009–1022
52. Maddock JR, Shapiro L (1993) Polar location of the chemoreceptor complex in the *Escherichia coli* cell. Science 259:1717–1723
53. Lybarger SR, Maddock JR (1999) Clustering of the chemoreceptor complex in *Escherichia coli* is independent of the methyltransferase CheR and the methylesterase CheB. J Bacteriol 181:5527–5529
54. Skidmore JM, Ellefson DD, McNamara BP, Couto MM, Wolfe AJ, Maddock JR (2000) Polar clustering of the chemoreceptor complex in *Escherichia coli* occurs in the absence of complete CheA function. J Bacteriol 182:967–973
55. Umehara S, Inoue I, Wakamoto Y, Yasuda K (2007) Origin of individuality of two daughter cells during the division process examined by the simultaneous measurement of growth and swimming property using an on-chip single-cell cultivation system. Biophys J 93(3):1061–1067
56. Stock, JB, Surette MG (1996) Chemotaxis. In Neidhardt FC, Curtiss R III, Ingraham JL, Lin ECC, Low KB, Magasanik B, Rfznikopp WS, Riley M, Schaechter M, Umbarger HE (eds) *Escherichia coli* and *Salmonella*: cellular and molecular biology, 2nd edn. ASM, Washington, pp 1103–1129
57. Magariyama, Y, Sugiyama S, Muramoto K, Kawagishi I, Imae Y, Kudo S (1995) Simultaneous measurement of bacterial flagellar rotation rate and swimming speed. Biophys J 69:2154–2162
58. Amsler CD, Cho M, Matsumura P (1993) Multiple factors underlying the maximum motility of *Escherichia coli* as cultures enter post-exponential growth. J Bacteriol 175:6238–6244
59. Pruss BM, Matsumura P (1996) A regulator of the flagellar regulon of *Escherichia coli*, *flhD*, also affects cell division. J Bacteriol 178:668–674
60. Aizawa SI, Kubori T (1998) Bacterial flagellation and cell division. Genes Cells 3:625–634
61. Hattori A, Umehara S, Wakamoto Y, Yasuda K (2003) Measurement of incident angle dependence of swimming bacterium reflection using on-chip single-cell cultivation assay. Jpn J Appl Phys 2 42:L873–L875
62. Umehara S, Hattori A, Wakamoto Y, Yasuda K (2004) Simultaneous measurement of growth and movement of cells exploiting on-chip single-cell cultivation assay. Jpn J Appl Phys 1 43:1214–1217
63. Wakamoto Y, Umehara S, Matsumura K, Inoue I, Yasuda K (2003) Development of non-destructive, non-contact single-cell based differential cell assay using on-chip microcultivation and optical tweezers. Sens Actuators B Chem 96:693–700
64. Mesibov R, Adler J (1972) Chemotaxis toward amino acids in *Escherichia coli*. J Bacteriol 112:315–326
65. Alon U, Camarena L, Surette MG, Aguera y Arcas B, Liu Y, Leibler S, Stock JB (1998) Response regulator output in bacterial chemotaxis. EMBO J 17:4238–4248
66. Maki N, Gestwicki JE, Lake EM, Kiessling LL, Adler J (2000) Motility and chemotaxis of filamentous cells of *Escherichia coli*. J Bacteriol 182:4337–4342
67. Wakamoto Y, Yasuda K (2006) Epigenetic inheritance of elongated phenotypes between generations revealed by individual-cell-based direct observation. Meas Sci Technol 17:3171–3177

68. Cohn M, Horibata K (1959) Physiology of the inhibition by glucose of the induced synthesis of the β-galactoside-enzyme system of *Escherichia Coli.* J Bacteriol 78:601–12
69. Pal C, Miklos I (1999) Epigenetic inheritance, genetic assimilation and speciation. J Theor Biol 200:19–37
70. Jablonka E, Lachmann M, Lamb MJ (1992) Evidence, mechanisms and models of the inheritance of acquired characters. J Theor Biol 158:245–268
71. Jablonka E, Lamb MJ (1998) Epigenetic inheritance in evolution. J Evol Biol 11:159–183
72. Mameli M (2004) Nongenetic selection and nongenetic inheritance. Br J Philos Sci 55:35–71
73. Solomon F (1981) Specification of cell morphology by endogenous determinants. J Cell Biol 90:547–553
74. Casadesus J, D'Ari R (2002) Memory in bacteria and phage. Bioessays 24:512–518
75. Gardner TS, Cantor CR, Collins JJ (2000) Construction of a genetic toggle switch in *Escherichia coli.* Nature 403:339–42
76. Ozbudak EM, Thattai M, Lim HN, Shraiman BI, van Oudenaarden A (2004) Multistability in the lactose utilization network of *Escherichia coli.* Nature 427:737–40
77. Wakamoto Y, Umehara S, Matsumura K, Inoue I, Yasuda K (2003) Development of non-destructive, non-contact single-cell based differential cell assay using on-chip microcultivation and optical tweezers. Sens Actuators B Chem 96:693–700
78. Wakamoto Y, Ramsden J, Yasuda K (2005) Single-cell growth and division dynamics showing epigenetic correlations. Analyst 130:311–317
79. Mukaihara T, Enomoto M (1997) Deletion formation between the two Salmonella typhimurium flagellin genes encoded on the mini F plasmid: *Escherichia coli* ssb alleles enhance deletion rates and change hot-spot preference for deletion endpoints. Genetics 145:563–72
80. Yu XC, Margolin W (1999) FtsZ ring clusters in min and partition mutants: role of both the Min system and the nucleoid in regulating FtsZ ring localization. Mol Microbiol 32:315–326
81. Woldringh CL, Mulder E, Valkenburg JA, Wientjes FB, Zaritsky A, Nanninga N (1990) Role of the nucleoid in toporegulation of division. Res Microbiol 141:39–49
82. de Boer PA, Crossley RE, Rothfield LI (1989) A division inhibitor and a topological specificity factor coded for by the minicell locus determine proper placement of the division septum in *E. coli.* Cell 56:641–9
83. de Boer PA, Crossley RE, Rothfield LI (1992) Roles of MinC and MinD in the site-specific septation block mediated by the MinCDE system of *Escherichia coli.* J Bacteriol 174:63–70
84. Bi E, Lutkenhaus J (1993) Cell division inhibitors SulA and minCD prevent formation of the FtsZ ring. J Bacteriol 175:1118–25
85. Raskin DM, de Boer PA (1999) Rapid pole-to-pole oscillation of a protein required for directing division to the middle of *Escherichia coli.* Proc Natl Acad Sci USA 96:4971–4976
86. Raskin DM, de Boer PA (1999) MinDE-dependent pole-to-pole oscillation of division inhibitor MinC in *Escherichia coli.* J Bacteriol 181:6419–24
87. Hu Z, Lutkenhaus J (1999) Topological regulation of cell division in *Escherichia coli* involves rapid pole to pole oscillation of the division inhibitor MinC under the control of MinD and MinE. Mol Microbiol 34:82–90
88. Fu X, Shih YL, Zhang Y, Rothfield LI (2001) The MinE ring required for proper placement of the … its cellular location during the *Escherichia coli* division cycle. Proc Natl Acad Sci USA 98:980–985
89. Meinhardt H, de Boer PA (2001) Pattern formation in *Escherichia coli*: a model for the pole-to-pole oscillations of Min proteins and the localization of the division site. Proc Natl Acad Sci USA 98:14202–14207
90. Pichoff S, Lutkenhaus J (2001) *Escherichia coli* division inhibitor MinCD blocks septation by preventing Z-ring formation. J Bacteriol 183:6630–6635
91. Howard M, Rutenberg AD, de Vet S (2001) Dynamic compartmentalization of bacteria: accurate division in *E. coli.* Phys Rev Lett 87:278102

92. Huang KC, Meir Y, Wingreen NS (2003) Dynamic structures in *Escherichia coli*: spontaneous formation of MinE rings and MinD polar zones. Proc Natl Acad Sci USA 100:12724–12728
93. Hiraga S, Niki H, Ogura T, Ichinose C, Mori H, Ezaki B, Jaffe A (1989) Chromosome partitioning in *Escherichia coli*: novel mutants producing anucleate cells. J Bacteriol 171:1496–1505
94. Aarsman, ME, Piette A, Fraipont C, Vinkenvleugel TM, Nguyen-Disteche M, den Blaauwen T (2005) Maturation of the *Escherichia coli* divisome occurs in two steps. Mol Microbiol 55:1631–1645
95. Den Blaauwen T, Buddelmeijer N, Aarsman ME, Hameete CM, Nanninga N (1999) Timing of FtsZ assembly in *Escherichia coli*. J Bacteriol 181:5167–5175
96. Thomas R (1998) Laws for the dynamics of regulatory circuits. Int J Dev Biol 42:479–485
97. Thomas R, Kaufman M (2001) Multistationarity, the basis of cell differentiation and memory. II. Logical analysis of regulatory networks in terms of feedback circuits. Chaos 11:180–195
98. Hernday AD, Braaten BA, Low DA (2003) The mechanism by which DNA adenine methylase and PapI activate the pap epigenetic switch. Mol Cell 12:947–957
99. van der Woude M, Braaten B, Low D (1996) Epigenetic phase variation of the pap operon in *Escherichia coli*. Trends Microbiol 4:5–9
100. McNairn AJ, Gilbert DM (2003) Epigenomic replication: linking epigenetics to DNA replication. Bioessays 25:647–656

Adv Biochem Engin/Biotechnol (2011) 124: 83–98
DOI: 10.1007/10_2010_84

Published Online: 12 November 2010

Measuring the Mechanical Properties of Single Microbial Cells

Colin R. Thomas, John D. Stenson and Zhibing Zhang

Abstract Many cells are considered to be susceptible to mechanical forces or "shear" in bioprocessing, leading to undesirable cell breakage or adverse metabolic effects. However, cell breakage is the aim of some processing operations, in particular high-pressure homogenisation and other cell disruption methods. In either case, the exact mechanisms of damage or disruption are obscure. One reason for this is that the mechanical properties of the cells are generally unknown, which makes investigation or prediction of the damage difficult. There are several methods for measuring the mechanical properties of single microbial cells, and these are reviewed briefly. In the context of bioprocessing research, a powerful method of characterising the mechanical properties of single cells is compression testing using micromanipulation, supplemented by mathematical modelling of the cell behaviour in compression. The method and associated modelling are described, with results mainly from studies on yeast cells. Continuing difficulties in making a priori predictions of cell breakage in processing are identified. In future, compression testing by micromanipulation might also be used in conjunction with other single cell analytical techniques to study mechanisms controlling form, growth and division of cells and their consequential mechanical behaviour. It ought to be possible to relate cell wall mechanics to cell wall composition and structure, and eventually to underlying gene expression, allowing much greater understanding and control of the cell mechanical properties.

Keywords Single cell · Biomechanics · Micromanipulation

Abbreviations

AFM	Atomic force microscopy
ESEM	Environmental scanning electron microscope

C. R. Thomas (✉), J. D. Stenson and Z. Zhang
School of Chemical Engineering, University of Birmingham, Edgbaston,
Birmingham, B15 2TT, UK
e-mail: c.r.thomas@bham.ac.uk

J. D. Stenson
Center for Bioprocess Engineering Research, Department of Chemical Engineering,
University of Cape Town, Rondebosch, 7701, South Africa

FEA Finite element analysis
RT-PCR Reverse transcription polymerase chain reaction
TEM Transmission electron microscope

Contents

1 Introduction

As this review will describe, techniques have become available over the last decade that can be used to measure the mechanical properties of single microbial cells. In some cases, it has been possible to use mathematical models of cell mechanical behaviour to deduce key material properties of cells, and in particular of their walls. Clearly, there must be reasons for undertaking such research. Early investigations of the mechanical properties of single cells were focussed on problems in bioprocessing, in particular possible damaging effects of "shear" on animal cells [43, 45]. However, cells with walls such as bacteria and yeast cells are generally considered to be strong, and the main issue with such cells in bioprocessing is the requirement in some circumstances for deliberate cell disruption, for example in a high pressure homogeniser, in order to release cell contents for subsequent recovery [10]. It has been suggested that in a high pressure homogeniser, two mechanisms may cause cell disruption: fluid mechanical stresses in the valve unit and subsequent jet impingement on the impact ring [22]. However, the extent of disruption not only depends on the hydrodynamics, but also on the mechanical properties of the cells [15, 35], and only by characterising the mechanical properties of single cells, and the material properties of their walls, will it be possible to determine successfully the specific mechanisms of cell disruption (which may be different for different cell types). Furthermore, it has been suggested that there can be significant adventitious disruption of *Escherichia coli* in centrifugation [7], and knowledge of cell mechanical properties would inform understanding of mechanisms here too.

In most bacteria shape is maintained by the cell wall [14], and this will also be the case for other microbial cells. Another reason for investigating the mechanical properties of cells and cell wall material properties is to try to relate those properties to the cell wall composition and structure and eventually through physiology and metabolism to gene expression. In this way we might increase our understanding of the form of cells and their mechanisms of growth.

2 Methods to Measure Single Cell Mechanical Properties

Lim et al. [19] reviewed experimental techniques for single human cell and single molecule biomechanics, but did not cover work on microbial cells. A more recent review considered the mechanical characterisation of single particles, including alternative techniques to compression testing by micromanipulation [42]. The scope of this review was greater than just single microbial cells as it included animal and plant cell research and studies on non-biological particles such as tablets and microcapsules. Not all of the methods described are applicable to microbial cells, which are generally smaller and more robust than many of the other particles found in biological and functional products. The pressure probe technique is used to study water relations in plant cells [36], but can also provide information on their bulk elastic moduli. However, it cannot be used on cells of sizes much smaller than 50 μm. Micropipette aspiration [19] is widely used to characterise the mechanical properties of many types of animal cells, but only works with such large and easily deformed cells.

2.1 Osmotic Pressure Variation

Microbial cell volume changes with the external osmotic pressure of the suspending solution, and this can be used to infer some cell mechanical properties. The cells are suspended in hypotonic solution, which is changed to hypertonic, causing water to flow from the cells because of the osmotic pressure change. The dependence of cell volume on the external osmotic pressure can be determined using a Coulter Counter® or Malvern Mastersizer® [21, 30] or for single cells by image analysis [4]. In the latter case, the cells can be fixed to the surface of a visualisation chamber, using chitosan in the case of yeast cells [4], so that an individual cell can be observed despite the flow of solution through the chamber. It is possible to use the cell volume-external osmotic pressure relationship to find the internal (turgor) pressure-volume relationship, from which a bulk elastic modulus for the cell can be found [21]. In the case of exponentially growing *Saccharomyces cerevisiae* cells, this was about 3 MPa. However, a bulk modulus treats the cell as a homogeneous body rather than a liquid-filled shell (i.e., cytoplasm within a cell wall), and in any case it is only found indirectly. On the other hand, this is a good method for studying (yeast) cell water relations [9] provided the time responses of the mixing chamber and of the cells are considered. This technique is not able to cause microbial cell failure, which is important in disruption studies.

2.2 Atomic Force Microscopy

Atomic force microscopy (AFM) [5] is a powerful technique for imaging the topography of surfaces, including those of microbial cells and other surfaces. As a nano-indentation system (i.e., a method by which a nanometer scale probe can be pushed into a surface), it can give local values of cell wall mechanical properties by providing curves of force versus indentation distance. It can do this with piconewton sensitivity and nanometer lateral resolution. The basis of the method is well known, and the reader is referred to an excellent review of microbial applications by Gaboriaud and Dufrene [12]. The method primarily gives the cell spring constant, treating a cell as a linearly elastic spring [3, 12], but it is possible to estimate a cell wall elastic modulus using either a Hertz model of compression of the surface [12, 37] or finite element analysis [46].

A particularly interesting application of AFM to microbial cells is given by Touhami et al. [37], who measured the elastic (Young's) modulus of normal cell walls of a brewer's yeast (*Saccharomyces cerevisiae*) and that of a bud scar where one expects a high level of chitin. The bud scar had a modulus value some ten times higher than the value for the surrounding wall, i.e., it was significantly stiffer. This was an important observation, distinguishing regions of different elasticity resulting from cell division.

Atomic force microscopy is so sensitive it has even been used to demonstrate local mechanical motions of the cell wall of *Saccharomyces cerevisiae* [25]. These motions were periodic (frequencies around 1 kHz with amplitudes of about 3 nm) and depended on an active metabolism. It was postulated that the motions were generated by large-scale forces resulting from the action of many molecular motor proteins working in concert. This is a remarkable observation that shows the power of AFM.

Although AFM is now an indispensable tool for investigating single cell biomechanics, it is not able to exert enough force to cause cell wall failure, i.e. cell disruption. It is also probable that the method gives elastic moduli that are not only laterally local, but also primarily reflect wall surface properties. For example, Touhami et al. [37] used indentations of about 40 nm, applying a Hertz surface compression model to extract moduli. The yeast cell wall is about 100 nm thick [32] and multilayered [16, 18]. The outer layer of mannoproteins can be 50% of the mass of the wall [16], and it seems plausible that the elastic moduli quoted by

Touhami et al. [37] are for this outer layer rather than the β 1,3-glucan layer, which is claimed to be load bearing [47]. This problem might be overcome by using finite element analysis of the cell wall behaviour, as applied (to *Aspergillus nidulans*) by Zhao et al. [46]. It is probably significant that the moduli found by Zhao et al. [46] were more than an order of magnitude greater than those estimated by Touhami et al. [37] and were more comparable with the values found by compression testing on yeast cells (Sect. 2.4). In any case, it is difficult to see how local properties from AFM could be related to useful global values for prediction of cell disruption behaviour.

2.3 Optical Tweezers

The optical trapping method or "optical tweezers" uses a strongly focussed beam of laser light to trap and manipulate particles suspended in a medium. The basis of the method is well known, and the reader is referred to a short review by Grier [13]. In most nanomechanical applications, silica microbeads are attached to opposite sides of the cell. One bead is fixed to a surface, and the other is moved by the optical tweezers to stretch the cell, with forces in the piconewton range. Although there are many examples of this technique being used to study the mechanical properties of cells, these are usually animal cells, without cell walls and therefore of low stiffness. Unfortunately, such small forces are not sufficient for characterising the mechanical properties of cells with walls, at least for large deformations (e.g. to cell wall failure).

2.4 Compression Testing by Micromanipulation

2.4.1 Introduction to Compression Testing

Compression testing by micromanipulation is a technique by which a single cell can be compressed or stretched and the force-deformation behaviour determined. This technique was developed over a decade ago by two of the authors (C.R. Thomas and Z. Zhang) and has since been applied to characterising the mechanical properties of a wide range of microscopic particles including single-celled (usually suspension-cultured) microorganisms [42]. In particular, the method has been applied to *Escherichia coli* cells [28], *Saccharomyces cerevisiae* [20, 31, 32], hybridomas [44, 45], tomato fruit cells [39] and tomato suspension-cultured cells [6]. Work on *Saccharomyces cerevisiae* is described more fully in Sect. 2.4.4 and a study on *Escherichia coli* in Sect. 2.4.5.

It should be noted that Peeters et al. [24] proposed similar equipment to study the biomechanical response of skeletal muscle cells, but it does not appear to have been used on microbial cells.

2.4.2 Experimental Setup

The method of compression testing by micromanipulation is described in Blewett et al. [6] and Wang et al. [39]. Figure 1 shows a photograph of the most important parts of the equipment used by these workers and by Stenson et al. [34].

A cell in suspension is compressed between a glass fibre probe and the bottom of the chamber containing the suspension. The compression surfaces represent two flat parallel surfaces that may be considered infinite with respect to the cell. Figure 2 shows an image of a yeast cell prior to compression between the probe and chamber.

The probe is mounted on a force transducer capable of measuring forces of 1 μN and greater. The extent of compression of the cell is known as a micromanipulator is used to drive the probe down onto the cell. In this way force-displacement data can be generated, which can be converted into force-cell deformation curves. The compression speed of the equipment was up to 68 μms^{-1}. Recent improvements using a piezo-electric stack to move the stage and cell against the probe allow speeds of at least 1.5 mm s^{-1} to be achieved [38]. The advantage of high speeds is that compression of cells can be rapid enough that water loss or other time-dependent effects can be neglected in modelling [34, 39].

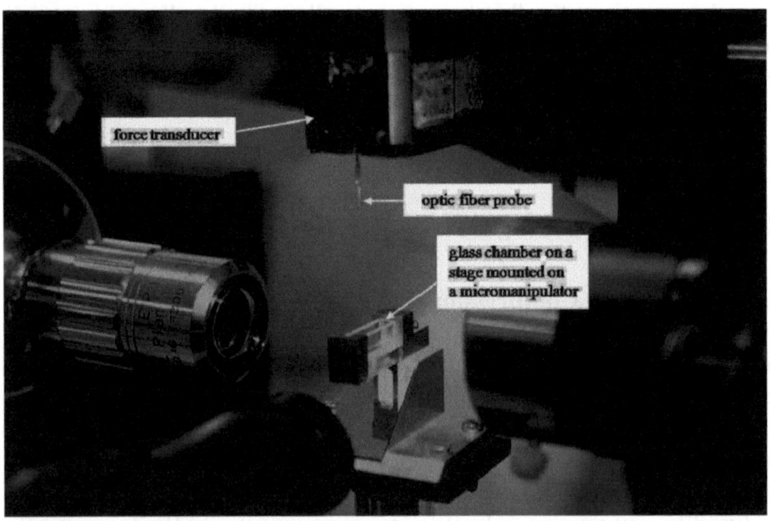

Fig. 1 Photograph of the high strain rate compression testing equipment. A sample of cell suspension is placed in the glass chamber, where the cell under test can be seen from beneath using an inverted microscope, and from the side using a second microscope. The glass probe attached to the force transducer is positioned over the cell. The stage carrying the glass chamber is driven upwards by the micromanipulator. As the cell is compressed, force displacement data are generated. The low strain rate compression tester is similarly arranged except the force transducer and glass probe are moved downwards by the micromanipulator, compressing the cell against the bottom of the chamber

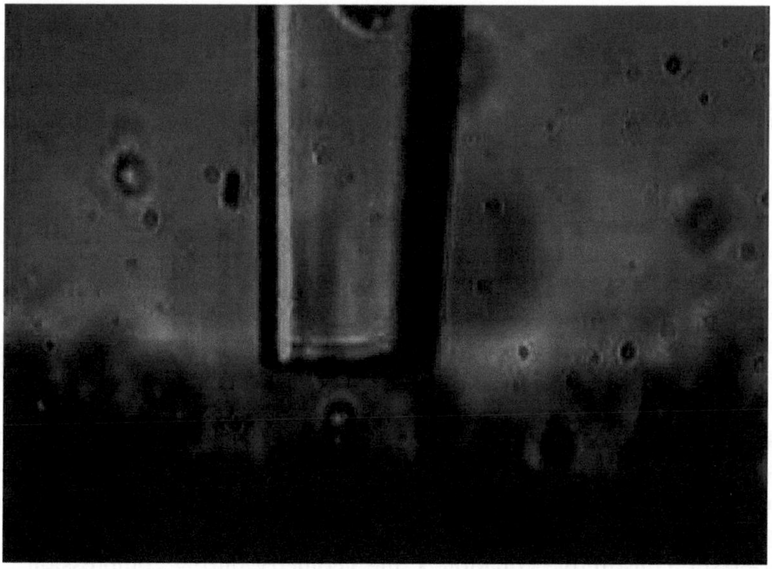

Fig. 2 *Saccharomyces cerevisiae* cell positioned under a glass probe ready for compression. In this image the probe base is 30 μm in diameter, and the yeast cell is ca. 5 μm in diameter; magnification: ×350

Fig. 3 Force displacement curve for a yeast cell. Initial cell diameter 5.0 μm. Compression speed 68 μms^{-1}

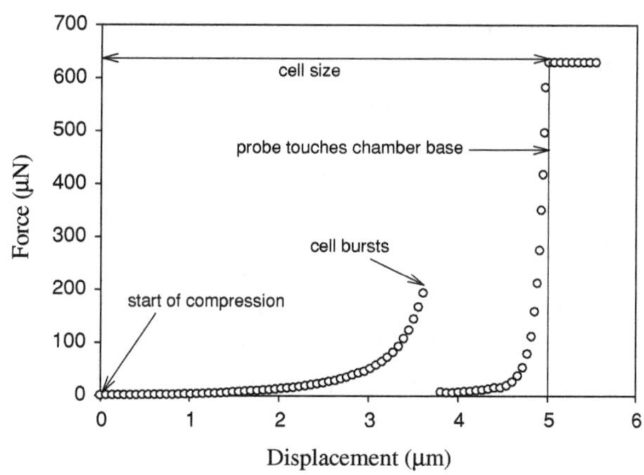

Figure 3 shows a dimensional force-displacement plot for a dried baker's yeast, resuspended in buffer, and compressed at 68 μms^{-1}. The cell size was 5 μm, so compression to bursting at 66% deformation took about 49 ms.

Unfortunately, force at wall failure (cell bursting) is not an intrinsic material property of the cell or its wall, and it would be better to have properties like the wall elastic modulus or the stresses in the wall at failure. Conceivably, these could be used in research into mechanisms of cell disruption or other cell wall studies. For this purpose, mathematical modelling is needed, as described in the following section.

2.4.3 Mathematical Modelling of Compression

The purpose of mathematical modelling of force-displacement data from compression testing is to extract cell wall material properties. The earliest work on this problem was by Feng and Yang [11] and Lardner and Pujara [17], who derived the governing equations for compression of a thin elastic spherical shell, gas or liquid-filled, respectively. These equations related local wall stresses to the corresponding strains as the shell was compressed. In this way theoretical force-deformation curves could be generated for given choices of elastic modulus and wall thickness. The modelling is difficult; even if the wall material can be assumed to be homogeneous and isotropic and to obey a version of Hooke's law (having a "linear-elastic constitutive equation"), the geometry during compression is very complicated, and the pressure inside the shell and the shape change during the compression. Nevertheless, approximating microbial cells as water-filled spheres with thin elastic walls has been the basis of much subsequent modelling, for example of yeast cells by Smith et al. [29–31]. More recently, an improved model has been developed that allows for the high strains and associated rotations that can occur during large compressions, for example to cell bursting [34]. This new model was based on work-conjugate Kirchoff stresses and Hencky strains, improving on the earlier use of infinitesimal strains. It is worth noting that there has been a recent detailed analysis of the general problem of the contact of a spherical membrane enclosing a fluid with rigid parallel plates [23], but this work has not yet influenced modelling of cell compression.

The assumptions of all of these and other models are that cells are initially spherical, that they have a relatively thin elastic wall surrounding an incompressible fluid, and that each single cell is compressed between two parallel surfaces, effectively of infinite size. In general, the cell is inflated at the start; its internal (turgor) pressure is greater than that outside. This matters because the wall is pre-stressed before compression, and this must be taken into account. The ratio of the inflated diameter to the original diameter is called the "initial stretch ratio" and is an important modelling parameter, along with the elastic modulus.

There is another important issue to be considered, which is that the cell wall/membrane of microbial cells is usually porous. This means that water can flow from a cell during compression, reducing the turgor pressure and consequently the stresses and strains in the wall. Although these water flows can be modelled, an additional parameter (the cell wall permeability) is required, and this is hard to identify [29]. For this reason it is helpful to have a high rate of compression, so any water flows during compression can be neglected. This was the approach taken by Stenson et al. [34] for compression testing of yeast cells.

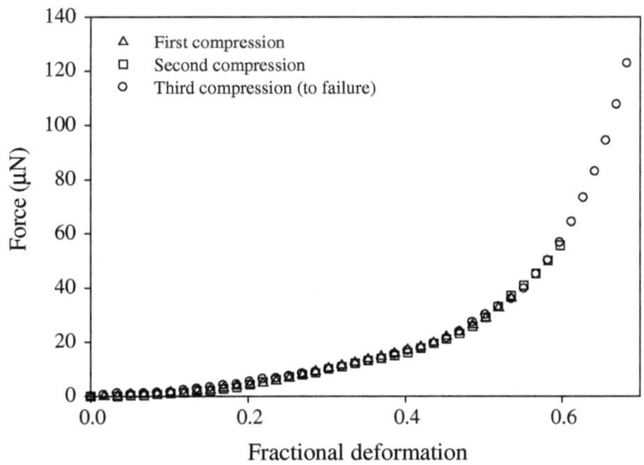

Fig. 4 Repeated compression of a single *Saccharomyces cerevisiae* cell, indicating that the cell wall is elastic up to at least 60% deformation and probably up to failure. Cell size = 4.1 μm, compression speed 68 $\mu m s^{-1}$, deformation at failure = 68%, force at failure = 123 μN

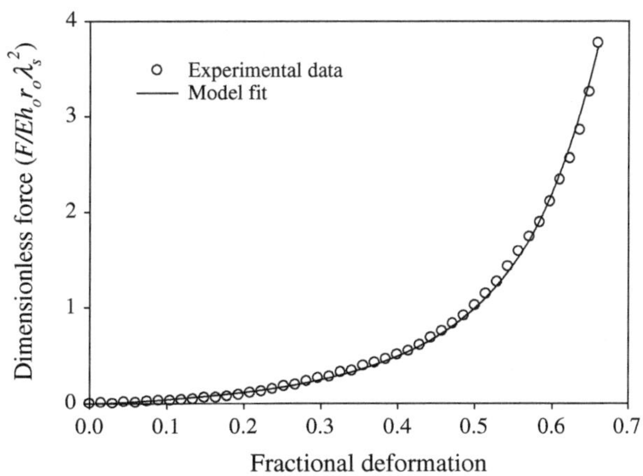

Fig. 5 Typical fit of the model of [34] to data from compression of a single *Saccharomyces cerevisiae* cell. Cell diameter 4.6 μm, estimated cell wall thickness 84 nm, best fit elastic modulus 172 MPa, best fit initial stretch ratio 1.04. Correlation coefficient (R^2) 0.998

Clearly the assumption of wall linear elasticity, especially across the whole range of cell deformation to bursting, may be false. Indeed, the cell walls of suspension-cultured tomato cells have been found to be elastic-plastic, showing irreversible deformations at higher strains [40]. Fortunately, the wall of a baker's yeast strain has been shown to be linearly elastic to failure [34]. Figure 4 shows how a test cell was repeatedly compressed and then released. Up to a deformation of about 60%, the curves overlapped, suggesting elasticity to this extent at least. Model fits (Fig. 5) showed that this was probably the case up to failure and that a linear-elastic wall constitutive equation could be applied throughout. This linear elasticity of the wall is of course a global description that does not discriminate between regions of different mechanical properties, in particular bud scars (see Sect. 2.2).

Whatever the extent of the elastic region, the elastic modulus and the initial stretch ratio may be found by matching experimental and modelling data until the best fit is determined. In the case of yeast cells, once an appropriate fit to cell bursting has been found, estimates can be made of wall stresses and strains at failure.

An alternative approach to analytical modelling is to use finite element analysis (FEA). This has the advantage that it is relatively easy to allow explicitly for water loss and elastic-plastic cell wall behaviour or other wall material constitutive equations. FEA was used by Smith et al. [29] for modelling compression of yeast cells. As analytical models become impractical as descriptions of the cell behaviour become more complex (e.g. elastic–plastic wall behaviour, local variations in wall properties), it is probable that FEA will eventually replace analytical modelling.

2.4.4 *Saccharomyces cerevisiae* Studies

Using compression testing and the model of Stenson et al. [34], intrinsic cell wall mechanical properties of a strain of baker's yeast (rehydrated from the dried state) have been determined. Because water loss could be neglected, it was possible to use modelling to obtain an elastic modulus, the initial stretch ratio, and (maximum) stresses, strains and strain energy per unit volume at failure. The mean values were 185 ± 15 MPa, 1.039 ± 0.006, 115 ± 5 MPa, 0.46 ± 0.03 and 30 ± 3 MPa (or MJm^{-3}) respectively [32], although it must be emphasised that data were obtained for single cells in this population; this is a single cell technique. It is possible the values will be less for freshly grown yeast cells, rather than rehydrated, previously dried, cells and this is currently being investigated. The elastic modulus might be compared with typical values for rubber (ca. 10–100 MPa), structural steel (ca. 200 GPa) and (suspension-cultured) tomato cell walls at low strain (2.3 ± 0.2 GPa; Wang et al. [40]). The strain energy per unit volume at failure may be useful in cell disruption studies, on the assumption that any cell breakage in processing would require the energy stored in the cell wall to exceed some critical level. The energy in cell walls might be estimated by mathematical modelling of how cells interact with fluid flow in processing equipment.

It appeared from the study of Stenson [32] that the elastic modulus and initial stretch ratio were essentially independent of cell size, whilst showing substantial biological variability (wide range of values for cells of similar sizes). However, all the failure criteria decreased with increasing cell size. It is possible that the elastic modulus and initial stretch ratio were not affected significantly by the presence of a few bud scars in the cell walls (Fig. 6), and any effect was masked by the variability, whilst the scars caused stress concentrations in the walls. The assumption of the model that the walls are homogeneous may be questionable. Because compression testing is a single-cell technique, it should be possible to investigate this further. See Sect. 3. Figure 6 also illustrates that assuming yeast cells are spheres is only an approximation. This matters as lack of sphericity implies that the walls vary in material properties around the cell, possibly not only due to bud scars. Fortunately, yeast cells are generally close to spherical, and the

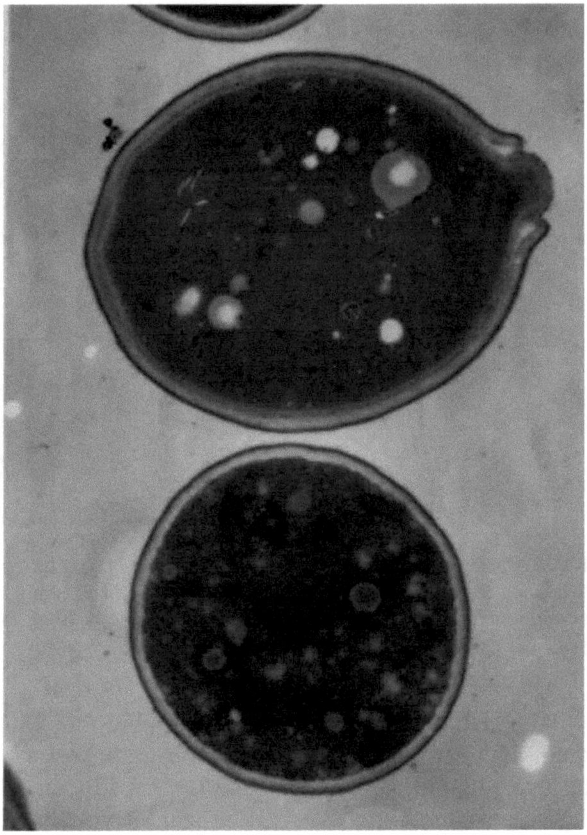

Fig. 6 TEM image of yeast cells, one showing a bud scar in cross section. Magnification: ×10,000

approximation has to be accepted at present because techniques for modelling non-spherical cells are not yet available.

It should also be noted that the elastic modulus that is found is dependent on the value chosen for the cell wall thickness, and at present this cannot be measured for a single cell as it is being compressed. The wall thickness is estimated from the cell radius using a pre-determined (mean) cell wall thickness to cell radius ratio, taken from transmission electron microscopy images. In any case, the assumption that the cell wall is isotopic is incorrect as it is multi-layered. If the load-bearing layer is only the β 1,3-glucan layer [47], then it may be possible to estimate the elastic modulus of this layer and relate it to the composition and structure of this layer.

2.4.5 Other Microbial Single-Cell Studies

Compression testing by micromanipulation has been applied to some bacterial cells [28]. Figure 7 shows an image of an *Escherichia coli* cell trapped between the surface of a probe and the chamber containing the cell suspension.

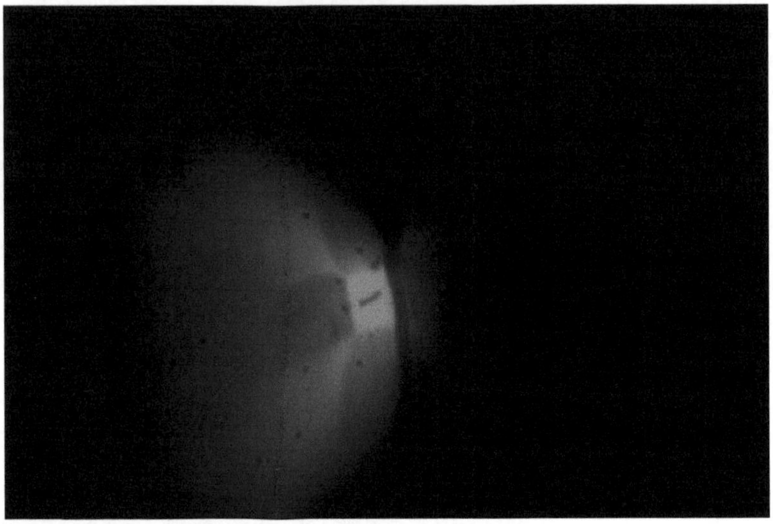

Fig. 7 *Escherichia coli* cell on the surface of an optical fibre probe as seen through the base of the chamber of the compression testing apparatus. The illuminated flat region, ground down from a 50-μm optic fibre probe, was ca. 7 × 7 μm. Photo: C Shiu. Reprinted from Thomas and Zhang [35] with permission

Although the glass probes are optic fibres, they were not chosen originally for their light-carrying properties. However, bacteria are so small that achieving adequate lighting for visualisation is very difficult. This was achieved by passing light down the optic fibre to the cell. It was found that the bursting force of *Escherichia coli* was about 4 μN. However, it has not yet been possible to model non-spherical bacteria like *E. coli* so intrinsic cell wall material properties are not yet available by compression testing.

2.5 Nanomanipulation

It has been shown that an environmental scanning electron microscope (ESEM)-based nanomanipulation technique is able to measure the force imposed on single yeast cells during compression between two parallel surfaces to determine their deformation including accurate cell shape information [26, 33]. Unfortunately, beam damage may be an issue, even for ESEM. Nevertheless, this approach offers new opportunities for gathering force-deformation data not only for yeast cells, but also for bacteria, which are often of sizes at the limit of the capability of conventional micromanipulation.

Ahmad et al. [1, 2] penetrated the cell walls of yeast cells using a nanomanipulator integrated into an ESEM. The penetration force rose with cell size and increased on average as the culture moved from log to stationary phase.

Elastic (Young's) moduli of about 3 MPa were found, although these values relate to the stiffness of the whole cell rather than just the wall. It will be useful to use compression testing by micromanipulation (Sect. 2.4) to study yeast cell wall material properties over the time course of a batch fermentation in the hope of understanding why penetration force rose with cell size and fermentation time.

3 Future Prospects for Compression Testing

Work is in hand to use compression testing to characterise the mean intrinsic mechanical properties of growing yeast cells throughout fermentations. However, as compression testing is a single cell technique, it opens up the prospect of using it in conjunction with other single cell analytical techniques to investigate these properties for sub-populations of cells. For example, flow cytometry might be used to sort cells by number of bud scars or other physiological factors to see how the wall properties vary with cell age or condition. It should also be possible to strip off cell wall layers enzymatically or chemically to confirm or disprove the claim that the β 1,3-glucan layer is the load-bearing layer of the yeast cell wall. If this layer bears the load, it should be possible to estimate its stiffness and to discover if this can be predicted from a molecular model of the layer. These methods might also be applied to cell wall mutants in a search for links between cell wall composition and structure and mechanical properties, and to investigate the mechanisms controlling form, growth and division of cells and their consequential mechanical behaviour. The ability to measure wall properties of single cells will remove some of the adverse effects of population heterogeneity that hinder other research in this area.

Finite element analysis (FEA) will be essential if intrinsic material properties are to be derived from compression testing data from cells with elastic-plastic wall behaviour or cells with inhomogeneous cell walls such as yeast cells. More importantly, FEA will be essential for work on non-spherical cells such as *E. coli*. However, the nanomanipulation methods need to be developed further for this to be really useful, as optical microscopy only gives crude shape information for such small cells.

AFM (Sect. 2.2) is a powerful method for obtaining local mechanical properties. Such data might be input into appropriate FEA models to model whole cells with scars. However, the differences in elastic moduli between AFM and compression testing by micromanipulation suggest that some reconciliation is required here, for example by discovering whether AFM is only a surface technique in this context, or whether the modelling of compression data is in some way flawed.

With good data on cell mechanical properties, it should be possible to investigate the mechanisms of cell damage or deliberate disruption in processing, for example in high pressure homogenisation or centrifugal recovery. There is the interesting possibility of predicting cell breakage by estimating the energy imparted to cells by fluid flow in processing equipment and comparing that with the strain energy per unit volume required for cell walls to failure.

Finally, compression testing might also be used to investigate cell genetic and metabolic responses to mechanical stresses, for example using RT-PCR. This has already been achieved with chondrocytes, chondrons and encapsulated chondrocytes [41] and should be possible with microbial cells. In the yeast case, the composition and architecture of the wall is controlled by about 1,200 genes [8]. Some obvious targets for directly manipulating the cell wall composition and structure to investigate consequential effects on the material properties are deletion of mcd4 leading to an increase in β 1,6-glucan levels and a reduction in GPI-anchored protein and mannan levels, cwh41Δ, which is very low in β 1,6-glucans, and gas1Δ, a cell wall assembly mutant [27].

4 Conclusion

There are several single-cell techniques for measuring the mechanical properties of single microbial cells. For investigations of local variations around the wall, atomic force microscopy is a powerful tool. If ambiguity about what exactly is being measured by nano-indentation can be resolved, then this will be the method of choice for such local measurements. The data could then be used in finite element analysis to develop more sophisticated models of microbial cell mechanics than those presently available. It may also be possible to use atomic force microscopy for cell wall penetration to obtain information about wall failure. However, modelling of this needs further development and validation. At present, the authors believe that compression testing by micro- or nano-manipulation provides the most scope for future research into cell wall material properties up to failure, which will be required if the mechanisms of cell disruption and cell damage in processing are to be understood. Furthermore, as a single-cell technique rather than a sub-cell technique, compression testing will be particularly suited to investigating the gene expression and cell physiology that underlie cell mechanical properties. Conceivably it will be feasible to combine the method with other single-cell techniques such as fluorescent microscopy in order to obtain simultaneously data on material properties and cell physiology. An alternative approach would be to use flow cytometry to sort cells by physiological or genetic markers before compression testing. Compression testing will be particularly valuable if analytical models are supplemented or replaced by finite element analysis, allowing cells with a wider range of sizes and shapes to be studied.

Acknowledgments The authors acknowledge the work of many colleagues, particularly Mr. C. Cowen of Micro Instruments (Oxford) Ltd. for equipment development and Professor A. Middelberg, Dr. A.E. Smith and Dr. P. Hartley for developments in modelling. Dr. C. Shiu is thanked for the photograph in Fig. 7. The authors also acknowledge the financial support of the Engineering and Physical Sciences Research Council, the Biotechnology and Biological Sciences Research Council and others.

References

1. Ahmad MR, Nakajima M, Kojima S et al (2008a) In situ single cell mechanics characterization of yeast cells using nanoneedles inside environmental SEM. IEEE T Nanotechnol 7:607–616
2. Ahmad MR, Nakajima M, Kojima S et al (2008b) The effects of cell sizes environmental conditions and growth phases on the strength of individual W303 yeast cells inside ESEM. IEEE T Nanobiosci 7:185–193
3. Arfsten J, Leupold S, Bradtmöller C et al (2010) Atomic force microscopy studies on the nanomechanical properties of *Saccharomyces cerevisiae*. Colloid Surf B 79:284–290
4. Berner JL, Gervais P (1994) A new visualization chamber to study the transient volumetric response of yeast-cells submitted to osmotic shifts. Biotechnol Bioeng 43:165–170
5. Binnig G, Quate CF, Gerber C (1986) Atomic force microscope. Phys Rev Lett 56:930–933
6. Blewett J, Burrows K, Thomas C (2000) A micromanipulation method to measure the mechanical properties of single tomato suspension cells. Biotechnol Lett 22: 1877–1883
7. Chan G, Booth AJ, Mannweiler K, Hoare M (2006) Ultra scale-down studies of the effect of flow and impact conditions during *E coli* cell processing. Biotechnol Bioeng 95:671–683
8. de Groot PWJ, Ruiz C, de Aldana CRV et al (2001) A genomic approach for the identification and classification of genes involved in cell wall formation and its regulation in *Saccharomyces cerevisiae*. Comp Funct Genome 2:124–142
9. de Maranon IM, Gervais P, Molin P (1997) Determination of cells' water membrane permeability: unexpected high osmotic permeability of *Saccharomyces cerevisiae*. Biotechnol Bioeng 56:62–70
10. Donsi F, Ferrari G, Lenza E, Maresca P (2009) Main factors regulating microbial inactivation by high-pressure homogenization: operating parameters and scale of operation. Chem Eng Sci 64:520–532
11. Feng WW, Yang WH (1973) Contact problem of an inflated spherical nonlinear membrane. J Appl Mech 40:209–214
12. Gaboriaud F, Dufrene YF (2007) Atomic force microscopy of microbial cells: application to nanomechanical properties, surface forces and molecular recognition forces. Colloid Surf B 54:10–19
13. Grier DG (2003) A revolution in optical manipulation. Nature 424:810–816
14. Harold FM (2007) Bacterial morphogenesis: learning how cells make cells. Curr Opin Microbiol 10:591–595
15. Kleinig AR, Middelberg APJ (1998) On the mechanism of microbial cell disruption in high-pressure homogenization. Chem Eng Sci 53:891–898
16. Klis FM, Boorsma A, De Groot PWJ (2006) Cell wall construction in *Saccharomyces cerevisiae*. Yeast 23:185–202
17. Lardner TJ, Pujara P (1980) Compression of spherical cells. Mech Today 5:161–176
18. Lesage G, Bussey H (2006) Cell wall assembly in *Saccharomyces cerevisiae*. Microbiol Mol Biol R 70:317–343
19. Lim CT, Zhou EH, Li A et al (2006) Experimental techniques for single cell and single molecule biomechanics. Mat Sci Eng C Biomim 26:1278–1288
20. Mashmoushy H, Zhang Z, Thomas CR (1998) Micromanipulation measurement of the mechanical properties of baker's yeast cells. Biotechnol Tech 12:925–929
21. Meikle AJ, Reed RH, Gadd GM (1988) Osmotic adjustment and the accumulation of organic solutes in whole cells and protoplasts of *Saccharomyces cerevisiae*. J Gen Microbiol 134:3049–3060
22. Moore EK, Hoare M, Dunnill P (1990) Disruption of baker's yeast in a high-pressure homogenizer—new evidence on mechanism. Enz Microb Technol 12:764–770
23. Nadler B (2010) On the contact of a spherical membrane enclosing a fluid with rigid parallel planes. Int J NonLinear Mech 45:294–300

24. Peeters EAG, Bouten CVC, Oomens CWJ, Baaijens FPT (2003) Monitoring the biomechanical response of individual cells under compression: a new compression device. Med Biol Eng Comput 41:498–503
25. Pelling AE, Sehati S, Gralla EB et al (2004) Local nanomechanical motion of the cell wall of *Saccharomyces cerevisiae.* Science 305:1147–1150
26. Ren YL, Donald AM, Zhang ZB (2008) Investigation of the morphology, viability and mechanical properties of yeast cells in Environmental SEM. Scanning 30:435–442
27. Sakai Y, Azuma M, Takada Y et al (2007) *Saccharomyces cerevisiae* mutant displaying beta-glucans on cell surface. J Biosci Bioeng 103:161–166
28. Shiu C, Zhang Z, Thomas CR (1999) A novel technique for the study of bacterial cell mechanical properties. Biotechnol Tech 13:707–713
29. Smith AE, Moxham KE, Middelberg APJ (1998) On uniquely determining cell-wall material properties with the compression experiment. Chem Eng Sci 53:3913–3922
30. Smith AE, Zhang Z, Thomas CR (2000a) Wall material properties of yeast cells: part 1. Cell measurements and compression experiments. Chem Eng Sci 55:2031–2041
31. Smith AE, Zhang ZB, Thomas CR et al (2000b) The mechanical properties of *Saccharomyces cerevisiae.* PNAS 97:9871–9874
32. Stenson JD (2008) Investigating the mechanical properties of yeast cells. PhD Thesis University of Birmingham, UK
33. Stenson JD, Ren Y, Donald AM, Zhang Z (2010) Compression testing by nanomanipulation in environmental scanning electron microscope. Exp Tech 34:60–62
34. Stenson JD, Thomas CR, Hartley P (2009) Modelling the mechanical properties of yeast cells. Chem Eng Sci 64:1892–1903
35. Thomas CR, Zhang Z (1998) The effect of hydrodynamics on biological materials. In: Galindo E, Ramirez OT (eds) Advances in bioprocess engineering II:137–170. Kluwer, London
36. Tomos D (2000) The plant cell pressure probe. Biotechnol Lett 22:437–442
37. Touhami A, Nysten B, Dufrene YF (2003) Nanoscale mapping of the elasticity of microbial cells by atomic force microscopy. Langmuir 19:4539–4543
38. Wang C, Cowen C, Zhang Z, Thomas CR (2005) High-speed compression of single alginate microspheres. Chem Eng Sci 60:6649–6657
39. Wang CX, Pritchard J, Thomas CR (2006) Investigation of the mechanics of single tomato fruit cells. J Texture Stud 37:597–606
40. Wang CX, Wang L, Thomas CR (2004) Modelling the mechanical properties of single suspension-cultured tomato cells. Ann Bot Lond 93:443–453
41. Wang QG, Magnay JL, Nguyen B et al (2009) Gene expression profiles of dynamically compressed single chondrocytes and chondrons. Biochem Biophl Res Commun 379:738–742
42. Zhang Z, Stenson JD, Thomas CR (2009) Micromanipulation in mechanical characterisation of single particles. In: Li J (ed) Advances in chemical engineering. Elsevier, Amsterdam, 37:29–85
43. Zhang Z, Al-Rubeai M, Thomas CR (1993) Estimation of disruption of animal cells by turbulent capillary flow. Biotechnol Bioeng 42:987–993
44. Zhang Z, Ferenczi MA, Lush AC, Thomas CR (1991) A novel micromanipulation technique for measuring the bursting strength of single mammalian cells. Appl Microbiol Biot 36:208–210
45. Zhang Z, Ferenczi MA, Thomas CR (1992) A micromanipulation technique with a theoretical cell model for determining mechanical properties of single mammalian cells. Chem Eng Sci 47:1347–1354
46. Zhao L, Schaefer D, Xu H et al (2005) Elastic properties of the cell wall of *Aspergillus nidulans* studied with atomic force microscopy. Biotechnol Prog 21:292–299
47. Zlotnik H, Fernandez MP, Bowers B, Cabib E (1984) *Saccharomyces cerevisiae* mannoproteins form an external cell-wall layer that determines wall porosity. J Bacteriol 159:1018–1026

Adv Biochem Engin/Biotechnol (2011) 124: 99–122
DOI: 10.1007/10_2010_96

Published Online: 12 November 2010

Single Cell Analytics: An Overview

Hendrik Kortmann, Lars M. Blank and Andreas Schmid

Abstract The research field of single cell analysis is rapidly expanding, driven by developments in flow cytometry, microscopy, lab-on-a-chip devices, and many other fields. The promises of these developments include deciphering cellular mechanisms and the quantification of cell-to-cell differences, ideally with spatio-temporal resolution. However, these promises are challenging as the analytical techniques have to cope with minute analyte amounts and concentrations. We formulate first these challenges and then present state-of-the-art analytical techniques available to investigate the different cellular hierarchies—from the genome to the phenome, i.e., the sum of all phenotypes.

Keywords Cell population heterogeneity · Envirostat · Omics technologies · Single cell analysis

Contents

H. Kortmann, L. M. Blank (✉) and A. Schmid (✉)
Laboratory of Chemical Biotechnology, Department of Biochemical and Chemical Engineering, TU Dortmund University, Emil-Figge-Str. 66, 44221 Dortmund, Germany
e-mail: lars.blank@bci.tu-dortmund.de

A. Schmid
Leibniz-Institut für Analytische Wissenschaften-ISAS-e.V., Otto-Hahn-Str. 6b, 44227 Dortmund, Germany
e-mail: andreas.schmid@bci.tu-dortmund.de

1 Size, Volume and Content of a Single Cell

Analytical challenges in single cell analysis are numerous, and mainly scale inversely with the size of the cell. Therefore, single cell analysis initially started with very large cells, such as the neurons from the Californian sea slug (*Aplysia californica*), one of the largest cells in nature (except bird eggs), with a volume of about 7 μl [1]. In contrast, the bacterium *Brevundimonas diminuta* has one of the smallest cells, with a diameter of 0.3 μm and a length of 1.0 μm. The volume is approximately 0.08 fl [2]. These two cell types differ in volume by a factor of about 1×10^8. Despite these extremes, cells used in many laboratories differ quite significantly in size. Three of the most commonly used organisms in biological research are *Escherichia coli* (bacterium), *Saccharomyces cerevisiae* (lower eukaryote), and human cells (higher eukaryote). *E. coli*, depending on the respective growth conditions, has a diameter of about 0.6 μm and a length of about 1.7 μm [2], the spherical *S. cerevisiae* a diameter of roughly 5 μm, and an average mammalian cell of about 10 μm. At first glance, these differences seem comparably minute. Another picture evolves when comparing the respective volumes, which are 0.5, 65, and 500 fl (Table 1). A *S. cerevisiae* cell is 140 times and a mammalian cell is 1,100 times larger in volume than an *E. coli* cell. The total

Table 1 Average cell content of a mammalian, a *S. cerevisiae*, and an *E. coli* cell

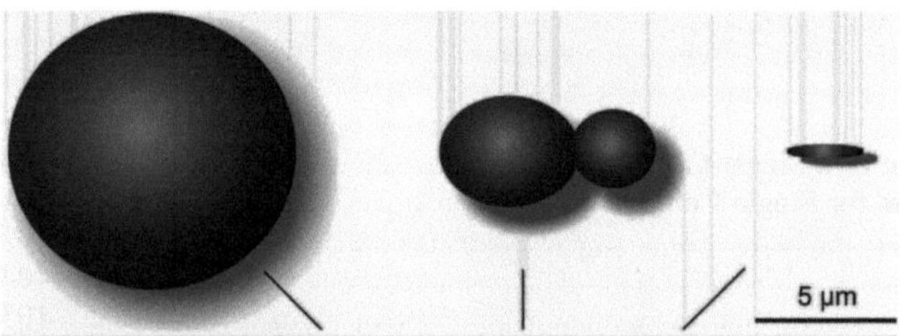

	Mammalian cell	*S. cerevisiae*	*E. coli*	
Diameter	10	5	2.5·0.7	μm
Volume	500	65	0.5	fL = μm³
Cell wet weight	$5 \cdot 10^5$	$7 \cdot 10^4$	$1 \cdot 10^3$	fg
Cell dry weight	$1 \cdot 10^5$	$2 \cdot 10^4$	$3 \cdot 10^2$	fg
Carbohydrates	$1 \cdot 10^4$	$8 \cdot 10^3$	$2 \cdot 10^1$	fg
DNA	$6 \cdot 10^3$	$2 \cdot 10^1$	5	fg
DNA	2	1-2[a]	1[b]	molecules
RNA	$2 \cdot 10^4$	$1 \cdot 10^3$	$6 \cdot 10^1$	fg
RNA	$3 \cdot 10^7$	$3 \cdot 10^6$	$1 \cdot 10^5$	molecules
Lipids	$1 \cdot 10^4$	$1 \cdot 10^3$	$3 \cdot 10^1$	fg
Lipids	$1 \cdot 10^{10}$	$1 \cdot 10^9$	$2 \cdot 10^6$	molecules
Protein (total)	$5 \cdot 10^4$	$9 \cdot 10^3$	$2 \cdot 10^2$	fg
Protein (total)	$1 \cdot 10^9$	$2 \cdot 10^8$	$4 \cdot 10^6$	molecules
Different proteins	$1 \cdot 10^4$	$4 \cdot 10^3$	$2 \cdot 10^3$	types
Protein (average)	5	2	$9 \cdot 10^{-2}$	fg
Protein (average)	$1 \cdot 10^5$	$5 \cdot 10^4$	$2 \cdot 10^3$	molecules

amount of analytes scales with the volume, and is therefore considerably lower in bacteria than in yeast and mammalian cells (Table 1).

The small volume of a single cell and the resulting extremely low absolute number of target molecules (Table 1) pose a great challenge for any analytical method used. It is obvious that the detection method must be very sensitive and capable of analyzing miniscule sample volumes. Alternatively, signal intensification might be necessary, e.g., by amplifying the analyte signal or by specifically labeling the analyte, allowing subsequent selective signal amplification. However, not only the performance of an analytical method sets the limit for the detection of compounds in single cells, but also the sample handling and its preparation. Manual handling of very low μl volumes can be challenging, but in single cell analysis nl and fl volumes have to be dealt with. Technologies like nanoHPLC and nanoESI as well as lab-on-a-chip (LOC) and μTAS technologies are methods of choice and good starting points for further developments [3]. For LOC technologies in particular, the construction of suitable interfaces between analytical instrumentation and microfluidic cell devices was and still is a major issue. Today, different analytical methods are available that enable single cell analysis.

Single cell analytics, as cell analytics, can be divided into cell consuming and non-consuming approaches. For example, optical methods and the analysis of cell products and environmental changes give evidence of the cellular status without consuming the cell. Other detection methods often require cell lysis, enabling research on genomic [4, 5], transcriptomic [6, 7], proteomic [8–10], and metabolomic levels [11, 12]. New methods allow non-cell-consuming research on these levels as well.

In the following, research on the level of a few to single cells is discussed in detail. After a short description of single cell analysis using microscopy, subsequent paragraphs are structured according to the cellular functional hierarchy into genome, transcriptome, proteome, and metabolome and cellular productivity analysis.

2 Microscopy, a Prerequisite for Single Cell Analysis

Analysis of the single cell requires knowledge of its existence. Knowledge of cells and microorganisms sounds commonplace today. However, Antoni van Leeuwenhoek is recognized as the first human ever to see microbes, reporting his discoveries from 1679 onwards [13, 14]. As late as 1838 the term bacteria was coined by Christian G. Ehrenberg [15]. These early works formed the foundation of the science of microbiology, founded mainly by Louis Pasteur, Ferdinand Cohn, and Robert Koch.

Microscopy is a standard technique today, with an often underestimated power. Yet, only the careful design of lenses, filters, and the light beam allows real-time analysis of single living cells. In fact, microbes, such as *E. coli*, are just above the physical resolution of a standard light microscope, which is ~0.2 μm.

With a light microscope, shape, size, volume, and motility of cells can be detected. When using a hemocytometer the concentration of cells in a given sample is quantifiable. Cell growth rates can be determined by long-time observations [16, 17]. More information is traceable using staining techniques. Stains, such as fluorophores, allow discriminating dead or live cells, for example, by staining dead cells, because of compromised membranes or are metabolized to fluorophores by living cells [18, 19]. Usage of fluorophore stains in microscopy enabled analysis of molecules and organelles at very low concentrations because of the high sensitivity of fluorescence detectors. Even metabolic activity can be directly measured in single cells using fluorescent substrates [20, 21]. A huge improvement in biological research by microscopy was the development of green fluorescent protein (GFP) and other fluorescent proteins as molecular tools. By tagging proteins with GFP, cell productivity [22], transport processes in cells [23], and spindle pole formation were revealed [16]. The latest biotechnological methods even allow designing fluorophores with different properties, such as emission and excitation spectra (e.g., GFP, eGFP, YFP), special fluorophores for anaerobic conditions [24], or to influence sensitivity to photo bleaching [25]. Careful selection of two fluorophores allowed, for example: (1) real-time detection of mRNA transcription and protein translation in single cells [26] or (2) the highly selective fluorescence energy transfer (FRET) technique [27]. At the end of this section, flow cytometry has to be mentioned. Flow cytometry is an optical high-throughput single cell analysis platform, with which cellular parameters like cell size and density can be measured. In addition, a large variety of fluorescence-based methods for single cell analysis exist. The high speed of flow cytometry (up to 10,000 nds of cells per second) allows the analysis of the heterogeneity of cell populations, but at the cost of spatiotemporal resolution in every single cell analyzed [28–30].

The techniques described in the section above use optical methods to gain cellular information. Despite the fact that these techniques make a wide variety of analytes and cellular processes traceable, they are limited to phenotypical analysis or require staining.

3 Genome

Genomic sequence data are the basis for every biological systems approach, as they harbor information about the capacity for metabolic activities, regulation, and finally growth and division of the cells of interest. For a deeper understanding of population heterogeneity, genomic to full genome analysis of each single cell would be ideal. The first advanced techniques allow DNA analysis of a few to single cells and are summarized here. Developments of genomic technologies on the few or single cell level are driven by research in preimplantation, prenatal, and forensic diagnostics [31–33]. Genomic single cell methods are often based on PCR [34, 35], utilizing its huge amplification capacity. The ultimate sensitivity is the amplification of a single DNA molecule followed by the detection of the many amplicons [36, 37]. Such high performance PCR methods were used to differentiate cells by the absence or presence of genetic targets [38]. First single cell PCR protocols were established with the inherent limitation to target only one or a small number of known sequences simultaneously [31].

This limitation can be overcome by whole genome amplification (WGA) techniques [39], which yield a DNA amount sufficient for additional analyses. Most WGA methods are also based on PCR technology, like DOP-PCR (degenerate oligonucleotide primed-PCR) [39, 40], PEP (primer extension preamplification) [39, 41], and LM-PCR (ligation-mediated PCR) [42]. Consequently, these methods share most intrinsic limitations of PCR, such as a relatively high error rate, allele drop out (ADO) [43], incomplete coverage, and a bias towards the percentage of GC content and loci. A potent alternative is the multiple displacement amplification (MDA) method. Advantages of MDA are its low amplification bias, high short tandem repeat (STR) and single nucleotide polymorphism (SNP) genotype completion rates, the lengths of resulting DNA products (>100,000 bp), and amplification rates [31, 43–47]. This non-PCR based WGA method is based on the highly processive bacteriophage Φ29 DNA polymerase [43, 48]. MDA was used to amplify genomic DNA from single mammalian cells [5, 49] and from single microbial cells [50, 51], such as a single *E. coli* cell [52]. The *E. coli* DNA was amplified 5 billion-fold, with a yield of several micrograms that allowed the sequencing of the multicopy 16S rRNA gene [52]. Heterogeneity in the genome copy number of human single cells from tissues was analyzed by combining whole genome amplification and high-resolution array comparative genomic hybridization (CGH) [53]. Single rat cancer cells were captured by laser microdissection, and of those cells, 122 genomic and mitochondrial loci were amplified by MDA. The amplified products were sequenced and analyzed by gel and capillary electrophoresis, revealing slippage and point mutations [54]. WGA techniques can thus close the gap between quantity and sensitivity, and are already in use to provide sufficient amounts of DNA for high throughput screening technologies such as SNP for personalized medicine applications [31, 55].

Alternative analytical techniques bridge the gap of quantity and sensitivity by decreasing the required DNA amount. Fluorescence methods are especially prominent when combined with microfluidic approaches. An example is laser-induced fluorescence (LIF), allowing the detection and quantification of DNA in the subpicomolar range [56], single-quantum-dot-based DNA nanosensors with a detection limit of about 50 DNA copies [57], and single DNA molecule detection (SMD) by molecular beacons [58]. In the ideal extreme, the idea of single molecule sequencing was experimentally demonstrated [59–61], which could make prior DNA amplification obsolete. So far, these methods are not applicable to bacteria, and no demonstration of the technology has been reported to the best of our knowledge.

Of highest importance is not only the detection method itself, but the sample preparation. Dilution effects during the process of DNA recovery, purification, and amplification were decreased by microfluidic strategies. For example, cells were lysed and PCR was performed in capillaries as well as in microfluidic chips [62, 63]. PCR methods were scaled down to nanoliter volumes [64, 65] and were improved by integrating cell isolation, cell lysis, DNA purification, and recovery on a nanoliter scale to analyze single mammalian and bacterial cells [32].

The above-described methods and developments already allow genetic research on the single cell level and will develop into routine whole single cell genome sequencing and analysis, with the prospect of screening whole cell populations. Assuming high enough analyte amounts, the current high throughput sequencing technologies (e.g., Illumina) can deliver in a one-week-long run up to 200 Gb of sequence information, providing 40,000 or more single bacterial genome sequences. When fully assembled genomes are desired, the development of single molecule sequencing technologies has to be followed (e.g., Helicos, Nanopore, Pacific Biosciences). When matured, these technologies might not only allow single genome sequencing without amplification, but also the detection of DNA modification, and hence can be used for epigenetic analyses of cell populations on the single cell level, adding another degree of complexity to cell function. The commercialization of new sequencing technologies just started a few years back; hence, major technological advances, including throughput, accuracy, read length, sensitivity, and price, can be expected.

4 Transcriptome

While genomic data provide insights into the stored information in organisms and the possible cellular functions, the transcriptome gives direct evidence of the information used by the cell. The transcriptome is the amount and composition of all messenger RNAs in a cell or a cell population. The abundance of the different mRNAs elucidates the status of the cell and can be a direct response to perturbation as in microbes mRNA half-lives are in the short minute range.

The reverse transcription (RT) of the RNA into the complementary DNA (cDNA) sequence enables transcriptome analysis with modified DNA techniques, such as RT-PCR. By using fluorescence-labeled primers, the amplification of the cDNA is quantifiable with RT-quantitative PCR (RT-QPCR, also called real-time PCR and denoted RT-PCR), allowing the determination of starting mRNA concentrations in a given sample. The well-established technologies RT-PCR and RT-QPCR enable transcriptome research of single cells with high accuracy and high sensitivity down to a single mRNA copy [66, 67]. The minute sample demand even allows parallel analysis of one cell [6, 68, 69]. With multiplex RT-QPCR, 5 transcripts [6, 70], 20 transcripts [68], and 30 transcripts [71] were measured from individual cells in parallel. It was found that the mRNA concentration of a specific gene differed up to nine-fold in single mouse cells [6] and up to 10,000-fold in monoclonal T cells [68].

Using RNA microarrays, wider screening for transcripts on the single cell level is feasible [72], although at the expense of sensitivity. Improvements of cDNA microarrays aim at intensifying fluorescence signals [73] or at better RNA amplification strategies [74, 75]. To get the needed probe amount, mRNA is amplified mainly by two methods. The first method is RT-PCR as described above. The second method is the amplification of antisense RNA (aRNA) by T7 RNA polymerase, which is less error prone and yields higher amplification yields of mRNA [76]. The transcriptome of individual pancreas cells was analyzed with microarrays. The detection limit was 25 transcript copies, prior to RT-PCR amplification, and a total of 95 separate transcripts were screened. Individual cells showed unpredicted gene expression combinations. Additionally, the finding that early pancreas consists of heterogeneous cell populations lead to a new definition of six different cell groups [77]. After T7-based linear amplification of mRNA samples, corresponding to 2 to 1,000 cell equivalents, 4,000 to 9,000 mRNAs were measured with a benchtop microarray and a lab-on-a-chip system [78]. Such analyses are increasingly parallelized [79]. The new parallelized and miniaturized sequencing technologies promise an unmatched resolution, with the possibility to measure the number of all mRNA molecules with in a single cell [80]. Although the reports are from higher cells, with the advantage of mRNAs with poly(A) tails for simple amplification starts, the development of microbial transcriptome analysis by direct sequencing is enforced and will be a widespread tool in the near future [81].

Single cell mRNA analysis techniques as described give more detailed information as averaged population results. However, they are still just a snapshot of a living system since they are cell-consuming techniques [82, 83]. Only tagging of mRNA in vivo provides spatiotemporal insight into a single cell [26, 84–86]. This was accomplished for mRNA concentration changes in a single cell [84] using a selective mRNA binding GFP fusion protein [87] and fluorescence correlation spectroscopy (FCS). Exposures of cells to steady levels of an inducer lead to an unexpected pulsating expression profile, which coincided with the cell cycle. This was contrary to the standard assumption gained by population research and is a landmark in single cell analysis. The average result obtained from 14 cells was equivalent to results from cell populations. With the same fusion protein, mRNA numbers were estimated [26, 88]. Quantal bursts were found even in fully induced cells (Fig. 1) [26]. In conclusion, the simultaneous use of different dyes in one cell was shown [26], and staining of more than one mRNA at a time was suggested [84]. This offers real-time analysis of gene networks on the single cell level.

5 Proteome

Until recently, standard proteomic methods required large sample quantities not obtainable from single cells [89]. Single cell proteome analysis necessitates specialized methods, geared for minute sample amounts, and equipped with extreme sensitivity. The typical protein content of differently sized single cells is given in Table 1. For example, assuming that a mammalian cell expresses 10,000 different

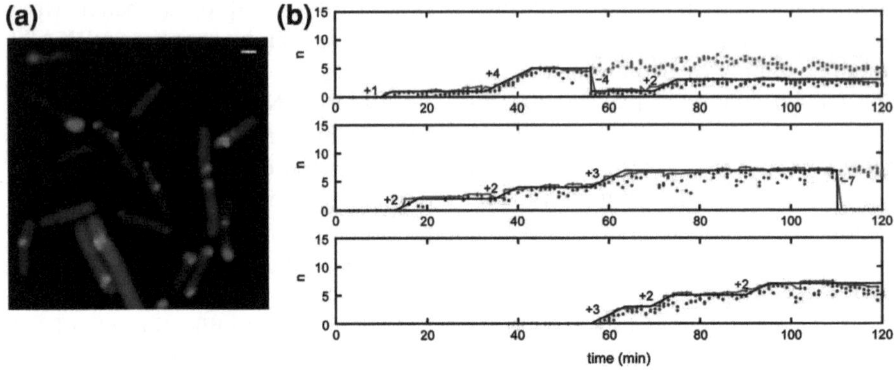

Fig. 1 Quantification of cell-to-cell differences. **a** Observation of mRNA (*green*, tagged by a GFP protein) and the translated *red* fluorescent protein in living *E. coli* cells. The picture is a false-colored overlay of the *green* and *red* channels. *Scale bar* 1 μm. **b** Estimated number of transcripts per cell, as a function of time. *Red dots* raw data, *green line* smoothed data, *black lines* fit by eye. The fit indicates periods of transcriptional inactivity and activity. Three out of four original diagrams were chosen. For further details, see text and original paper [26]. Reprinted from [26]. Copyright (2005), with permission from Elsevier

proteins, the average protein is present at the 200 zmol or 100,000 copies level [90].

Capillary electrophoresis (CE) is the method of choice for single cell proteome analysis, due to its minute sample demand and high separation efficiency. Single cells were analyzed by CE coupled with LIF (laser-induced fluorescence) detection. As a first step, the cell lysate or the single cell is brought into the capillary. Whole cells are lysed by chemicals [62], heat [91, 92], or electricity [93, 94]. Lysis of the cell in the capillary ensures that the complete cell content is in the capillary and circumvents unwanted dilution of the analytes in external preparation steps. In the next step, the cell content is separated by CE and often detected by LIF [95]. Detection by LIF requires native fluorescent proteins [96], proteins that are fused with a fluorescence tag [97, 98], marked due to derivatization [92] or tagged by fluorescent antibodies [99]. One drawback of CE-LIF is its inability to identify detected proteins directly [100]. This can be partly overcome by spiking the sample with known proteins (Fig. 2) [92].

The cell cycle dependence of protein expression was observed in a single cancer cell by protein fingerprinting, using capillary sieving electrophoresis with LIF detection [91]. After spiking the protein of interest, it was detectable in the single cell lysate [92]. In both experiments, the cell was injected into the capillary, then lysed, and the protein content was derivatized. Instead of unspecific derivatization, human interferon gamma was selectively labeled by fluorescent antibodies. The immunoreaction was performed in the capillary after lysis and products were detected by LIF, achieving a detection limit of 3.8 amol [99]. A further improvement was gained by in vivo (in the cell) binding of antibodies and proteins prior to lysis. Dilution and diffusion effects are decreased to the

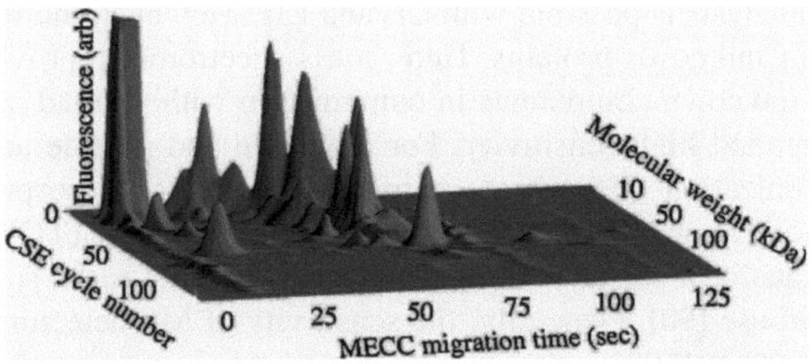

Fig. 2 Protein landscape image from single MC3T3-E1 cell. The figure presents the data in a landscape plot, where the height of the peak is proportional to the fluorescence intensity. Reprinted in part with permission from Hu et al. [8]. Copyright 2004 American Chemical Society

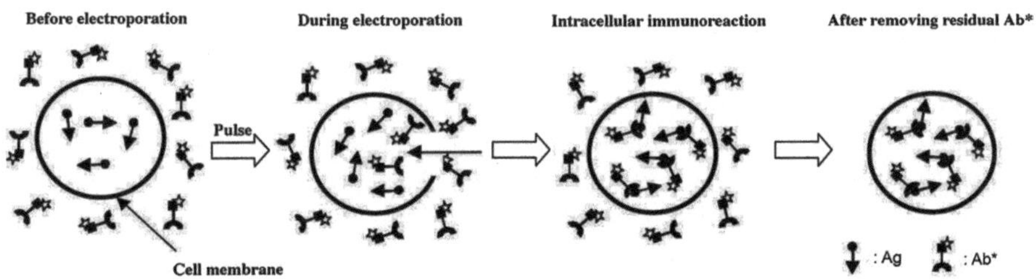

Fig. 3 The cell itself as reaction chamber; the schematic diagram shows the process of intracellular immunoreaction [101]. Reprinted from [101]. Copyright (2006), with permission from Elsevier

absolute minimum when using the cell as reaction chamber. The method resulted in a zeptomole detection limit when combined with CE-LIF (Fig. 3) [101]. A detection limit of 100 ymol for eGFP dilutions was reached in studies with CE-LIF; 100 ymol translates into 60 eGFP molecules. In cell experiments 20,000 eGFP molecules were measured after lysis of individual *Deinococcus radiodurans* cells [98]. Not by LIF, but by electrochemical detection after CE (CE-ECD) the glucose-6-phosphate dehydrogenase amount of erythrocytes was indirectly quantified by detecting the produced NADH, with a limit of detection of 1.3 zmol [102].

An even higher separation efficiency of complex mixtures is achieved by two-dimensional CE. Protein fingerprints of different single cells were obtained using capillary sieving electrophoresis/micellar electrokinetic capillary chromatography. Changes in protein abundance in single cells were detected for (1) osteoprogenitor cells (MC3T3-E1), native and recombinant, and (2) in native and apoptosis induced breast cancer cells (MC-F7). The fingerprints consisted in maximum of 167 local maxima after baseline correction. The detection sensitivity of the method is sufficient, while the protein separation efficiency has to be improved as it is an order of magnitude lower than in standard 2D SDS-PAGE [8].

Single cell analysis is possible with CE and LIF. However, the method fails for identification of unknown proteins. Here, mass spectrometry provides identification ability of unknown compounds in combination with a broad detection range, high specificity, and high sensitivity. For proteome and peptide analysis, mainly electrospray ionization (ESI) and matrix-assisted laser desorption ionization (MALDI) as ionization methods and ion trap, time of flight (TOF), quadrupole, and Fourier transform ion cyclotron resonance (FTICR) mass spectrometers as analyzers are in use [90]. Generally, the sensitivity of MS detectors is lower than that of LIF detectors [100].

Profiting from the huge cell size and thus absolute analyte number, MALDI-TOF analysis of single *A. californica* neurons with direct peptide sequencing after postsource decay was performed [103]. The spatial distribution of neuropeptides within single neurons was profiled with MALDI-MS. Neuropeptide amounts in the cells were in the femto- to picomole range [104]. By stable isotope labeling of a neuropeptide, quantitative measurements within single-neurons of the mollusks by MALDI-MS was doable [105]. In single rat cells, more than 15 signaling peptides were observed by MALDI-TOF. Moreover, 14 of these peptides were identified by comparison with population homogenates [106]. Shortly afterwards, an improved protocol of the method was published [107]. The proteome of phenotypically similar bacteria was recently analyzed by MS techniques after a heterogeneous cell population had been sorted using flow cytometry [108]. With such combinatorial techniques, sub-population analysis becomes truly feasible.

First demonstration of single cell analysis by CE-FTICR was performed on red blood cells detecting alpha-chains and beta-chains (each 450 amol cell^{-1}) of hemoglobin [109]. A few years later, hemoglobin chains were separated and detected by CE-ESI-TOF mass spectrometry [110]. In an approach to analyze the two chains plus carbonic anhydrase I and II (7 and 0.8 amol cell^{-1}), separation of the four components was achieved by improved CE. Whereas detection of the two chains was achieved from one cell, the detection limits of carbonic anhydrase I and II were 20 and 44 amol and were hence not detected [111]. With the coupling of ESI to FTICR, a sensitivity of 30 zmol for proteins ranging from 8 to 20 kDa was achieved [112]. Similar sensitivity was reached by means of nanoLC coupled online with a micro-solid-phase sample extraction and a nanoscale ESI interface to a FTICR-MS, detecting 75 zmol of individual proteins from 0.5 pg whole proteome extract [89], which might enable the analysis of single cells [113].

High throughput analysis of the influence of medium conditions on the proteome of single *S. cerevisiae* cells was applied to a library of >2,500 GFP-tagged yeast strains using flow cytometry. An automated setup measured seven samples per minute, with each sample counting >50,000 cells. The observed population heterogeneity on the proteome level differed especially under dynamic conditions from the known population heterogeneity on the transcriptome level [114]. In a similar approach, a cancer cell library with 1,200 different clones was analyzed. Time-lapse fluorescent microscopy enabled monitoring of dynamic proteome changes of individual cancer cells in response to a drug. In total, nearly 1,000 tagged proteins were observed during several days of cell growth. Undisturbed cells displayed a cell-to-cell fluorescence variability of 10–60% of the mean. In response to drug addition, the variability increased by about 30%. Notably, 24 proteins reacted highly differently in individual cells, from increasing over constant to decreasing concentrations inside the cell (Fig. 4) [115].

Microfluidic lab-on-a-chip platforms were used for proteome analysis. Improvements were made by integrating several preparation steps on a chip, thereby reducing dilution, delay, and dead volumes. For example, integrations of enrichment column, separation column, and nanoESI on polymer microfluidic chips were accomplished [116, 117], although they have not been evaluated yet for single cell analysis. On another chip, the protein contents of two different cell types, a human and an insect, were analyzed. In this study, single cells were transported to a chamber and then lysed. The lysis buffer contained fluorescently labeled antibodies, targeting specific proteins. After electrophoretic separation, the fluorescently tagged proteins were detected by confocal microscopy. The detection area of the microscope was enlarged by using cylindrical optics [9]. In another chip, single cell injection, high voltage lysis, derivatization, CE separation, and LIF detection were combined for the analysis of single red blood cells. The derivatized glutathione was detected at attomole concentrations [93]. A non-destructive method was shown, in which the release of neuropeptides from an *A. californica* neuron bag cell was collected on-chip by functionalized gold strips. Employment of three channels downstream of the cell allowed temporal analysis of protein release, since the channels were consecutively flushed for 15 min. The collected neuropeptides where then prepared off-line for and analyzed by MALDI-TOF, revealing temporal differences in neuropeptide release [118].

mRNA translation over complete cell live times was measured by fluorescent microscopy in different cell types. Heat shock-induced transcription and subsequently translation of eGFP tagged heat shock proteins in *S. cerevisiae* were quantified by measuring the overall cell fluorescence by microscopy. The cells were trapped in single cell traps on a chip [119]. In *E. coli* a membrane protein was conjugated with an YFP variant and transcribed under control of the *lac*-promoter, allowing the tracing of single proteins in single cells by microscopy. The cells were placed on an agarose gel [120]. In another survey, single cell gene induction dynamics were revealed, using an automated tracking and detection program for fluorescent photographs. A period of vulnerability to antibiotics in persister bacteria was thereby found. The *E. coli* cells were cultivated and analyzed in a lab-on-a-chip device with constant medium flow and changing inducer concentrations (Fig. 5) [121].

Especially for the fluorescence techniques, and for analysis of many proteins, the MS technologies will be used increasingly for single cell protein analysis. With respect to living single cell analysis, the analysis of secreted proteins seems to be a viable approach [122]. As the sensitivity and comprehensiveness of protein analysis including quantification was shown, and will be adapted by many groups, the dynamic range of the analytical method will move into focus. In bacteria, the difference between low and highly abundant proteins can be as high as 10^5, while current MS have a dynamic range of 10^3 to 10^4. Most likely, the relative change of

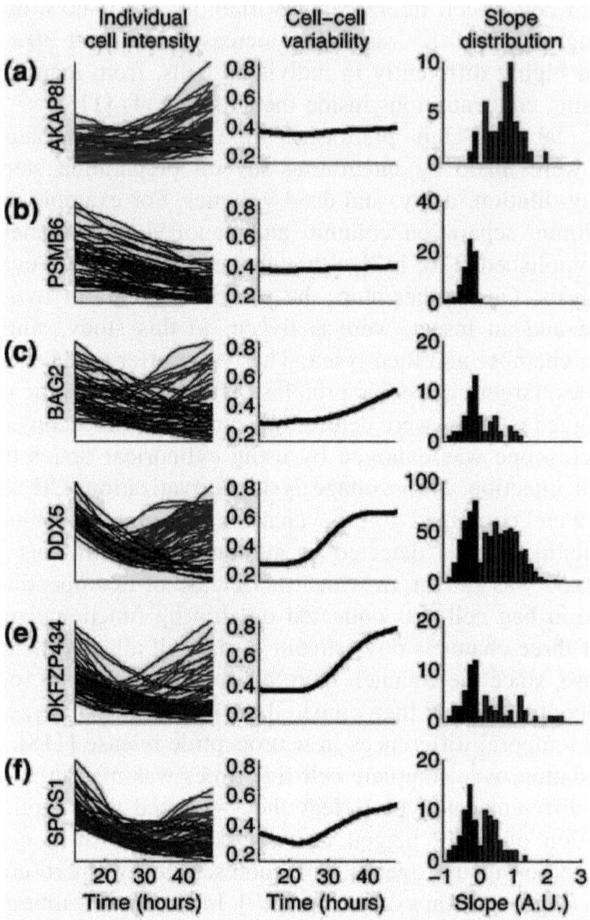

Fig. 4 A subset of proteins displays a bimodal response at the individual cell level in response to camptothecin (CPT). **a** and **b** Examples of proteins that do not show bimodal behavior, which are representative of most proteins in the study. Profiles are similarly shaped in each individual cell. Profiles rise with time (*red lines*) or decrease with time (*blue lines*) in parallel. Cell-cell variability (defined as standard deviation divided by the mean of cell-cell distribution at each time point) increases slightly over time, and the distribution of slopes of fluorescence levels shows uniform behavior. **c–f** Examples of proteins that show bimodal behavior. The dynamics after about 20 h vary between cells: Some cells show an increase in fluorescence levels (*red*) and other cells show a decrease (*blue*). This resulted in bimodal distributions of fluorescent intensity slopes measured in arbitrary units (A.U.). Slopes are defined as median temporal derivative of the fluorescence levels in the interval between 24 h after drug addition to 48 h (or time of cell death). Figure and text from [115]. Reprinted with permission from AAAS

a very rare transcription factor has more influence on the phenotype than a highly abundant ribosomal protein has. Therefore, dynamic range increases should go hand in hand with quantification accuracy allowing the researcher to trace regulatory decisions throughout the cellular networks.

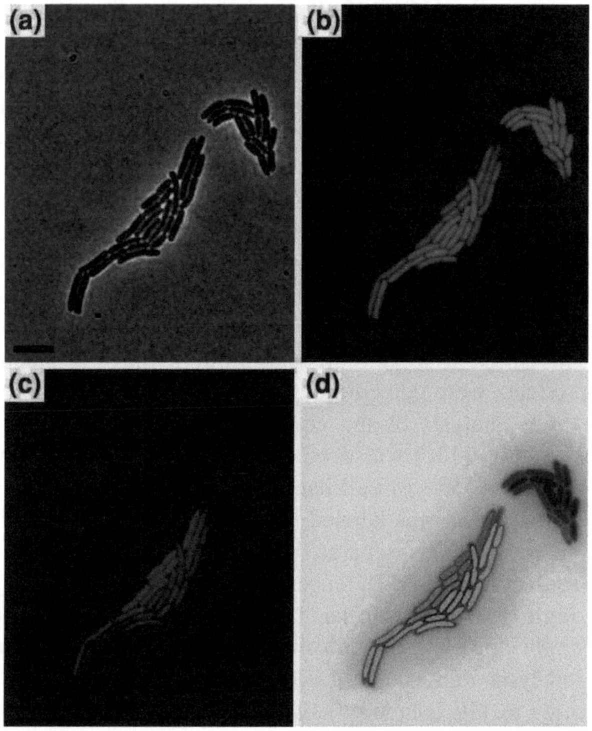

Fig. 5 The growth of *E. coli* microcolonies was monitored, which originated each from a single cell. **a–c** Simultaneous pictures of the same field of view, in **a** phase contrast; **b** YFP and **c** mCherry induced fluorescence. **d** Automated monitoring of growth and lineage; new cells are assigned an arbitrary color by the algorithm. *Scale bar* 5 μm. Figure from [121]. Copyright 2008 National Academy of Sciences, USA

6 Metabolic Activity and Metabolites

The metabolome comprises the sum of all small molecules produced by a cell. Metabolite concentrations and metabolite production rates give direct evidence of the fundamental cellular functions of transcription, translation, and biochemical performance, thus the present status of the cell. Metabolites typically have molecular weights below 1 kDa. When compared to the proteome, the additional analytical challenges are plenty. The metabolome has higher turnover rates, a by far larger chemical diversity, and a significantly lower mass content. Finally, the low molecular weights complicate analysis and detection by, for example, mass spectrometry [123]. Not only the metabolic content, but also the metabolic activity of a single cell is difficult to measure. Single cell production rates were estimated from population data: lysine production by bacteria is in the range of 0.2 fmol cell^{-1} s^{-1} [124], and the ethanol production rate is around 6 fmol cell^{-1} s^{-1} [125]. Therefore, a single yeast cell can produce about 1 ng ethanol in 1 h. Assuming that the yeast would be captured in a 0.1/μl microchamber, the ethanol concentration would be approximately 0.2 mM. In consequence, the sampling strategy and the sample volume are of high importance to minimize analyte loss and dilution.

Thus, a method of choice for metabolome analysis of single cells is CE. CE-LIF was employed to examine the correlation of cell cycle and metabolism in human cancer cells (HT29). Cells were fed on a fluorescent disaccharide and then lysed inside the capillary. The released biosynthetic and biodegraded products were separated and detected by LIF. Several cell subpopulations with unique biosynthetic patterns were identified [126]. Using a similar analytical method, glycosphingolipid metabolism in tumor cells (AtT-20) was monitored. Fluorescently tagged ganglioside was fed to the cells, which were fixed in formalin for 50 h, transferred to a CE capillary, lysed in the capillary, and metabolic products containing the fluorescent substrate were detected by LIF. The achieved detection limit was at the low zeptomol level [127]. Instead of feeding cells fluorogenic substrates, amino acids in erythrocytes were intracellularly derivatized. The derivatizing reagent was introduced into the living cells by electroporation. Thereafter, a single cell was brought into the electrode, lysed, and derivatized products were detected by LIF [128].

CE analysis of the contents of single mitochondria [129] and human cells (Jurkat T) [130] was demonstrated applying the lab-on-a-chip concept. The chips combined steps of cell handling and analyte separation. The acidic content of the mitochondria was labeled intracellularly. Mitochondria were then photolysed, and the content was separated by CE within milliseconds. The gained preliminary results indicated differences between individual cells [129]. All necessary steps for single human cell (Jurkat T) analysis by CE and LIF detection were combined on a single chip. The chip allowed single cell isolation, delivering of chemical reagents, cell lysis, and chemical derivatization, and as a last step, separation of the amino acids by CE coupled to LIF detection [130].

Sheath flow coupling of CE to ESI-MS enabled analysis of 6 nl aliquots of a single *A. californica* neuron cell, without prior derivatization of the analytes. The detection limit was in the low nanomole range for signaling molecules, such as acetylcholine, histamine, dopamine, and serotonine [131].

Only picoliters from single living cells were needed for live cell imaging of mammalian cells (rat, RBL-2H3). Cells were placed under a microscope, and with a nano-ESI tip, contents of the cytoplasma were withdrawn. The gained sample was then directly introduced via ESI into a MS. The method allowed spatial sampling of the metabolite contents of a cell. More then 700 peaks were detected, and the local distribution of histidine and tryptophan could be determined. With the knowledge of spatial occurrence of metabolites, metabolic processes, including compartimentation, might be identifiable [132]. A high-throughput MALDI method for chemical single cell analysis was developed and tested using *S. cerevisiae*. Samples were prepared by spotting picoliters on a thin matrix. Although not used for single cell analysis, the method has single cell sensitivity; its limit of detection was calculated to be 5 to 12 amol for species such as ADP, GDP, ATP, GTP, and acetyl-CoA [123]. Instead of spotting the samples, a microfluidic device was used for continuous sample deposition on the matrix. The continuous deposition of the analyte sample preserved the separation order [133]. Alternatively, a non-invasive technique for intracellular metabolite quantification based on proteins that respond with a conformational change to small molecule binding was developed by Frommer and co-workers. Using microscopy to quantify fluorescence resonance

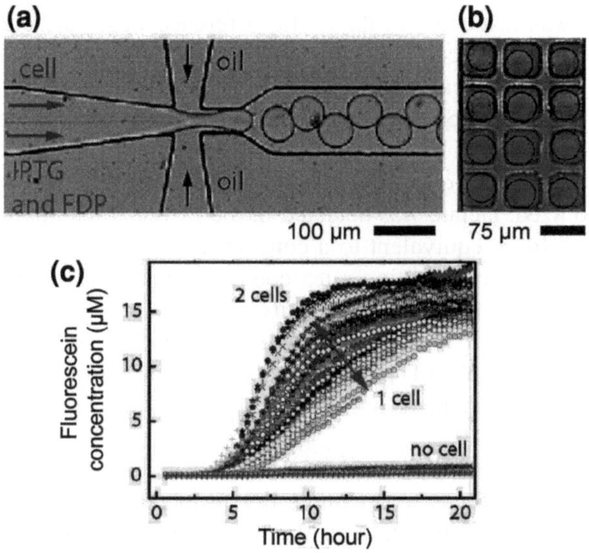

Fig. 6 Droplet-based single cell analysis. **a** Droplet-generating area of the microfluidic device. The cells and the solution containing substrate (FDP) and ITPG were combined on-chip before droplet formation. A typical droplet volume was approximately 20 pl. **b** Droplet storage area on the microfluidic chip, loaded with droplets. **c** Fluorescein concentration increase due to enzymatic conversion of FDP over time. Each series of symbols represents product formation in individual droplets starting with no cell (background hydrolysis), one or two cells in the droplet storage area. Figure composed of different original figures. Figures reprinted in part with permission from Shim et al. [20]. Copyright 2009 American Chemical Society

energy transfer (FRET) allows the detection of conformational changes, which can subsequently be correlated to metabolite concentrations [134]. The technique is also discussed for measuring intracellular reaction rates (fluxes) in single cells [135].

The analysis of metabolic conversion rates is only possible by spatiotemporal monitoring of a cell. Commonly, fluorogenic substrates are used, and the signal increase over time is quantified. Fluorescein diacetate (FDA) is a widely used substrate that when metabolized yields the fluorescent product fluorescein. The cellular changes were monitored in human cells (Jurkat T). From the fluorescein increase in microchambers, the numbers of present mRNAs and enzymes per *S. cerevisiae* cell were calculated [136]. The enzymatic activity of alkaline phosphatase per *E. coli* cell was spatiotemporally measured in a microdroplet lab-on-a-chip (Fig. 6) [20].

An enzymatic reaction is also detectable by electrochemical techniques. Lactate, for example, is quantifiable by amperometric detection of the amount of produced hydrogen peroxide when oxidized by lactate oxidase. In an ultra-low-volume microtiter chamber, lactate concentration changes of single living human heart cells were measured using two microelectrodes. A linear range from

65 to 266 fmol was achieved. A threefold difference in lactate content between healthy and anoxic single cells after metabolic inhibition demonstrated the value of the method for biological research [137]. With a lab-on-a-chip incorporating five electrodes, the lactate production of single heart cells was measured. One pair of the electrodes was used as pacing electrodes for field-stimulation of the cell. The three other electrodes quantified lactate production. When the heart cell was electro-permeabilized, lactate was detected in the extracellular space. The limit of detection was 4.8 fmol, equivalent to a concentration of 7.4 μM [138].

Single cell oxygen consumption rates could be measured by oxygen-sensitive platinum phosphor sensors in cell proximity. The sensors are excited by a laser, and changes in emission are recorded by a microscope, allowing calculating oxygen consumption [139–141]. With this measuring principle, oxygen consumption rates of single human lung cancer cells [139] and rat macrophage cells were measured [140, 141].

The analysis of metabolite concentrations and reaction rates in single cells will increasingly move into focus for example in cancer research. Metabolic network operation, i.e., the flux distribution in the metabolic network, is not sequenced based and is a proxy of the cellular phenotype [142]. Advances in existing and development of new analytical technologies to improve the reliability and usability of single cell metabolism analysis will be reported in the future.

7 Conclusion

Single cell analysis is feasible, allowing the monitoring of concentrations and concentration changes of certain analytes on all levels of the cellular hierarchy. Many methods provide, in principle, the sensitivity to detect low concentrations of analytes dissolved in buffer. However, it is generally difficult to reach the same sensitivity during single cell analysis, since femtoliter handling and dilution effects complicate the task. Especially low volume methods such as CE and lab-on-a-chips are advantageous [143–148]. However, due to the diversity and complexity of the cell content, none of the current methods enable comprehensive single cell analysis, and hence rather specific cellular processes, with very limited numbers of analytes are in the focus of research. To increase the portfolio of possible analytes, the development of analytical methods and interfaces to couple several analytical methods is desirable. As introduced, the analytical challenges scale not only with the number of cells, but also with the cell size. Many of today's robust and well-established single cell analytical tools have to be adapted in the future to conquer small unicellular organisms.

Although the challenges are huge, the rewards waiting to be gathered outweigh the required efforts. We are facing the genotype-phenotype conundrum, and hence can read and write the genetic code, but do not comprehend what we are reading nor do we know what to write. This conundrum is a result of a missing understanding of basic concepts of cellular function, including regulation on all levels of the cell hierarchy [149]. Only by verifying/falsifying hypotheses that address these basic cellular concepts, the existing gap in knowledge can be bridged. We are convinced that single cell analysis will contribute significantly to this development, as in single cell analysis the cellular complexity is not blurred by population averages. With the rapid developments in single cell analytics, we will see major contributions to our understanding of this cellular complexity.

References

1. Kreiner T, Kirk MD, Scheller RH (1987) Cellular and synaptic morphology of a feeding motor circuit in *Aplysia californica*. J Comp Neurol 264(3):311–325
2. Wang Y, Hammes F, Duggelin M, Egli T (2008) Influence of size, shape, and flexibility on bacterial passage through micropore membrane filters. Environ Sci Technol 42(17): 6749–6754
3. Schmid A, Kortmann H, Dittrich PS, Blank LM (2010) Chemical and biological single cell analysis. Curr Opin Biotechnol 21(1):12–20
4. Lange BM (2005) Single-cell genomics. Curr Opin Plant Biol 8(3):236–241
5. Jiang Z, Zhang X, Deka R, Jin L (2005) Genome amplification of single sperm using multiple displacement amplification. Nucleic Acids Res 33(10):e91
6. Bengtsson M, Stahlberg A, Rorsman P, Kubista M (2005) Gene expression profiling in single cells from the pancreatic islets of Langerhans reveals lognormal distribution of mRNA levels. Genome Res 15(10):1388–1392
7. Bengtsson M, Hemberg M, Rorsman P, Stahlberg A (2008) Quantification of mRNA in single cells and modelling of RT-qPCR induced noise. BMC Mol Biol 9:11
8. Hu S, Michels DA, Fazal MA, Ratisoontorn C, Cunningham ML, Dovichi NJ (2004) Capillary sieving electrophoresis/micellar electrokinetic capillary chromatography for two-dimensional protein fingerprinting of single mammalian cells. Anal Chem 76(14):4044–4049
9. Huang B, Wu H, Bhaya D, Grossman A, Granier S, Kobilka BK, Zare RN (2007) Counting low-copy number proteins in a single cell. Science 315(5808):81–84
10. Greif D, Galla L, Ros A, Anselmetti D (2008) Single cell analysis in full body quartz glass chips with native UV laser-induced fluorescence detection. J Chromatogr A 1206(1):83–88
11. Mapelli V, Olsson L, Nielsen J (2008) Metabolic footprinting in microbiology: methods and applications in functional genomics and biotechnology. Trends Biotechnol 26(9):490–497
12. van der Werf MJ, Overkamp KM, Muilwijk B, Coulier L, Hankemeier T (2007) Microbial metabolomics: toward a platform with full metabolome coverage. Anal Biochem 370(1):17–25
13. Ruestow EG (1983) Images and ideas: Leeuwenhoek's perception of the spermatozoa. J Hist Biol 16(2):185–224
14. Smit P, Heniger J (1975) Antoni van Leeuwenhoek (1632–1723) and the discovery of bacteria. Antonie van Leeuwenhoek 41(3):219–228
15. Siesser WG (1981) Christian Gottfried Ehrenberg: founder of micropaleontology. Centaurus 25(3):166–188
16. Hoepfner D, Brachat A, Philippsen P (2000) Time-lapse video microscopy analysis reveals astral microtubule detachment in the yeast spindle pole mutant cnm67. Mol Biol Cell 11(4):1197–1211
17. Kortmann H, Chasanis P, Blank LM, Franzke J, Kenig EY, Schmid A (2009) The Envirostat—a new bioreactor concept. Lab Chip 9(4):576–585
18. Soejima T, Iida K, Qin T, Taniai H, Yoshida S (2009) Discrimination of live, anti-tuberculosis agent-injured, and dead *Mycobacterium tuberculosis* using flow cytometry. FEMS Microbiol Lett 294(1):74–81

19. Strauber H, Muller S (2010) Viability states of bacteria-specific mechanisms of selected probes. Cytom A 77(7):623–634
20. Shim J-u, Olguin LF, Whyte G, Scott D, Babtie A, Abell C, Huck WTS, Hollfelder F (2009) Simultaneous determination of gene expression and enzymatic activity in individual bacterial cells in microdroplet compartments. J Am Chem Soc 131(42):15251–15256
21. Li PC, de Camprieu L, Cai J, Sangar M (2004) Transport, retention and fluorescent measurement of single biological cells studied in microfluidic chips. Lab Chip 4(3): 174–180
22. Schumann CA, Dorrenhaus A, Franzke J, Lampen P, Dittrich PS, Manz A, Roos PH (2008) Concomitant detection of CYP1A1 enzymatic activity and CYP1A1 protein in individual cells of a human urothelial cell line using a bilayer microfluidic device. Anal Bioanal Chem 392(6):1159–1166
23. Nyfeler B, Hauri HP (2007) Visualization of protein interactions inside the secretory pathway. Biochem Soc Trans 35(Pt 5):970–973
24. Drepper T, Eggert T, Circolone F, Heck A, Krauss U, Guterl JK, Wendorff M, Losi A, Gartner W, Jaeger KE (2007) Reporter proteins for in vivo fluorescence without oxygen. Nat Biotechnol 25(4):443–445
25. Shaner NC, Patterson GH, Davidson MW (2007) Advances in fluorescent protein technology. J Cell Sci 120(Pt 24):4247–4260
26. Golding I, Paulsson J, Zawilski SM, Cox EC (2005) Real-time kinetics of gene activity in individual bacteria. Cell 123(6):1025–1036
27. Greeson JN, Organ LE, Pereira FA, Raphael RM (2006) Assessment of prestin self-association using fluorescence resonance energy transfer. Brain Res 1091(1):140–150
28. Ibrahim SF, van den Engh G (2007) Flow cytometry and cell sorting. Adv Biochem Eng Biotechnol 106:19–39
29. Duhamel S, Gerardi G (2009) Detection of extracellular phosphatase activity at the single-cell level by enzyme-labeled fluorescence and flow cytometry: the importance of time kinetics in ELFA labeling. Cytom A 75A(2):163–168
30. Muller S, Nebe-von-Caron G (2010) Functional single-cell analyses: flow cytometry and cell sorting of microbial populations and communities. FEMS Microbiol Rev 34(4):554–587
31. Bergen AW, Haque KA, Qi Y, Beerman MB, Garcia-Closas M, Rothman N, Chanock SJ (2005) Comparison of yield and genotyping performance of multiple displacement amplification and OmniPlex whole genome amplified DNA generated from multiple DNA sources. Hum Mutat 26(3):262–270
32. Hong JW, Studer V, Hang G, Anderson WF, Quake SR (2004) A nanoliter-scale nucleic acid processor with parallel architecture. Nat Biotechnol 22(4):435–439
33. Hanson EK, Ballantyne J (2005) Whole genome amplification strategy for forensic genetic analysis using single or few cell equivalents of genomic DNA. Anal Biochem 346(2):246–257
34. Handyside AH, Pattinson JK, Penketh RJ, Delhanty JD, Winston RM, Tuddenham EG (1989) Biopsy of human preimplantation embryos and sexing by DNA amplification. Lancet 1 (8634):347–349
35. Fassihi H, Renwick PJ, Black C, McGrath JA (2006) Single cell PCR amplification of microsatellites flanking the COL7A1 gene and suitability for preimplantation genetic diagnosis of Hallopeau-Siemens recessive dystrophic epidermolysis bullosa. J Dermatol Sci 42(3):241–248
36. Lagally ET, Medintz I, Mathies RA (2001) Single-molecule DNA amplification and analysis in an integrated microfluidic device. Anal Chem 73(3):565–570
37. Rungpragayphan S, Kawarasaki Y, Imaeda T, Kohda K, Nakano H, Yamane T (2002) High-throughput, cloning-independent protein library construction by combining single-molecule DNA amplification with in vitro expression. J Mol Biol 318(2):395–405
38. Li H, Yeung ES (2002) Selective genotyping of individual cells by capillary polymerase chain reaction. Electrophoresis 23(19):3372–3380

39. Peng W, Takabayashi H, Ikawa K (2007) Whole genome amplification from single cells in preimplantation genetic diagnosis and prenatal diagnosis. Eur J Obstet Gynecol Reprod Biol 131(1):13–20
40. Cheung VG, Nelson SF (1996) Whole genome amplification using a degenerate oligonucleotide primer allows hundreds of genotypes to be performed on less than one nanogram of genomic DNA. Proc Natl Acad Sci USA 93(25):14676–14679
41. Dietmaier W, Hartmann A, Wallinger S, Heinmoller E, Kerner T, Endl E, Jauch KW, Hofstadter F, Ruschoff J (1999) Multiple mutation analyses in single tumor cells with improved whole genome amplification. Am J Pathol 154(1):83–95
42. Klein CA, Schmidt-Kittler O, Schardt JA, Pantel K, Speicher MR, Riethmuller G (1999) Comparative genomic hybridization, loss of heterozygosity, and DNA sequence analysis of single cells. Proc Natl Acad Sci USA 96(8):4494–4499
43. Dean FB, Hosono S, Fang L, Wu X, Faruqi AF, Bray-Ward P, Sun Z, Zong Q, Du Y, Du J, Driscoll M, Song W, Kingsmore SF, Egholm M, Lasken RS (2002) Comprehensive human genome amplification using multiple displacement amplification. Proc Natl Acad Sci USA 99(8):5261–5266
44. Hosono S, Faruqi AF, Dean FB, Du Y, Sun Z, Wu X, Du J, Kingsmore SF, Egholm M, Lasken RS (2003) Unbiased whole-genome amplification directly from clinical samples. Genome Res 13(5):954–964
45. Lasken RS, Egholm M (2003) Whole genome amplification: abundant supplies of DNA from precious samples or clinical specimens. Trends Biotechnol 21(12):531–535
46. Lasken RS (2007) Single-cell genomic sequencing using multiple displacement amplification. Curr Opin Microbiol 10(5):510–516
47. Hawkins TL, Detter JC, Richardson PM (2002) Whole genome amplification–applications and advances. Curr Opin Biotechnol 13(1):65–67
48. Panelli S, Damiani G, Espen L, Micheli G, Sgaramella V (2006) Towards the analysis of the genomes of single cells: further characterisation of the multiple displacement amplification. Gene 372:1–7
49. Ren Z, Zeng HT, Xu YW, Zhuang GL, Deng J, Zhang C, Zhou CQ (2009) Preimplantation genetic diagnosis for Duchenne muscular dystrophy by multiple displacement amplification. Fertil Steril 91(2):359–364
50. Lasken RS (2009) Genomic DNA amplification by the multiple displacement amplification (MDA) method. Biochem Soc Trans 37:450–453
51. Chang HW, Sung Y, Kim KH, Nam YD, Roh SW, Kim MS, Jeon CO, Bae JW (2008) Development of microbial genome-probing microarrays using digital multiple displacement amplification of uncultivated microbial single cells. Environ Sci Technol 42(16):6058–6064
52. Raghunathan A, Ferguson HR Jr, Bornarth CJ, Song W, Driscoll M, Lasken RS (2005) Genomic DNA amplification from a single bacterium. Appl Environ Microbiol 71(6): 3342–3347
53. Fiegler H, Geigl JB, Langer S, Rigler D, Porter K, Unger K, Carter NP, Speicher MR (2007) High resolution array-CGH analysis of single cells. Nucleic Acids Res 35(3):e15
54. Frumkin D, Wasserstrom A, Itzkovitz S, Harmelin A, Rechavi G, Shapiro E (2008) Amplification of multiple genomic loci from single cells isolated by laser micro-dissection of tissues. BMC Biotechnol 8:17
55. Church GM (2006) Genomes for all. Sci Am 294(1):46–54
56. Wang TH, Peng Y, Zhang C, Wong PK, Ho CM (2005) Single-molecule tracing on a fluidic microchip for quantitative detection of low-abundance nucleic acids. J Am Chem Soc 127(15):5354–5359
57. Zhang CY, Yeh HC, Kuroki MT, Wang TH (2005) Single-quantum-dot-based DNA nanosensor. Nat Mater 4(11):826–831
58. Zhang CY, Chao SY, Wang TH (2005) Comparative quantification of nucleic acids using single-molecule detection and molecular beacons. Analyst 130(4):483–488

59. Sauer M, Angerer B, Ankenbauer W, Foldes-Papp Z, Gobel F, Han KT, Rigler R, Schulz A, Wolfrum J, Zander C (2001) Single molecule DNA sequencing in submicrometer channels: state of the art and future prospects. J Biotechnol 86(3):181–201
60. Braslavsky I, Hebert B, Kartalov E, Quake SR (2003) Sequence information can be obtained from single DNA molecules. Proc Natl Acad Sci USA 100(7):3960–3964
61. Astier Y, Braha O, Bayley H (2006) Toward single molecule DNA sequencing: direct identification of ribonucleoside and deoxyribonucleoside 5′-monophosphates by using an engineered protein nanopore equipped with a molecular adapter. J Am Chem Soc 128(5):1705–1710
62. Irimia D, Tompkins RG, Toner M (2004) Single-cell chemical lysis in picoliter-scale closed volumes using a microfabricated device. Anal Chem 76(20):6137–6143
63. Cheng J, Sheldon EL, Wu L, Uribe A, Gerrue LO, Carrino J, Heller MJ, O'Connell JP (1998) Preparation and hybridization analysis of DNA/RNA from *E. coli* on microfabricated bioelectronic chips. Nat Biotechnol 16(6):541–546
64. Liu J, Enzelberger M, Quake S (2002) A nanoliter rotary device for polymerase chain reaction. Electrophoresis 23(10):1531–1536
65. He Y, Zhang YHH, Yeung ES (2001) Capillary-based fully integrated and automated system for nanoliter polymerase chain reaction analysis directly from cheek cells. J Chromatogr A 924(1–2):271–284
66. Hartshorn C, Rice JE, Wangh LJ (2003) Differential pattern of Xist RNA accumulation in single blastomeres isolated from 8-cell stage mouse embryos following laser zona drilling. Mol Reprod Dev 64(1):41–51
67. Stahlberg A, Bengtsson M (2010) Single-cell gene expression profiling using reverse transcription quantitative real-time PCR. Methods 50(4):282–288
68. Peixoto A, Monteiro M, Rocha B, Veiga-Fernandes H (2004) Quantification of multiple gene expression in individual cells. Genome Res 14(10A):1938–1947
69. Taniguchi K, Kajiyama T, Kambara H (2009) Quantitative analysis of gene expression in a single cell by qPCR. Nat Methods 6(7):503–506.
70. Diercks A, Kostner H, Ozinsky A (2009) Resolving cell population heterogeneity: real-time PCR for simultaneous multiplexed gene detection in multiple single-cell samples. PLoS ONE 4(7):e6326
71. Toledo-Rodriguez M, Blumenfeld B, Wu C, Luo J, Attali B, Goodman P, Markram H (2004) Correlation maps allow neuronal electrical properties to be predicted from single-cell gene expression profiles in rat neocortex. Cereb Cortex 14(12):1310–1327
72. Eberwine J, Kacharmina JE, Andrews C, Miyashiro K, McIntosh T, Becker K, Barrett T, Hinkle D, Dent G, Marciano P (2001) mRna expression analysis of tissue sections and single cells. J Neurosci 21(21):8310–8314
73. Xiang CC, Kozhich OA, Chen M, Inman JM, Phan QN, Chen Y, Brownstein MJ (2002) Amine-modified random primers to label probes for DNA microarrays. Nat Biotechnol 20(7):738–742
74. Zhou W, Abruzzese RV, Polejaeva I, Davis S, Davis S, Ji W (2005) Amplification of nanogram amounts of total RNA by the SMART-based PCR method for high-density oligonucleotide microarrays. Clin Chem 51(12):2354–2356
75. Jenson SD, Robetorye RS, Bohling SD, Schumacher JA, Morgan JW, Lim MS, Elenitoba-Johnson KS (2003) Validation of cDNA microarray gene expression data obtained from linearly amplified RNA. Mol Pathol 56(6):307–312
76. Hinkle D, Glanzer J, Sarabi A, Pajunen T, Zielinski J, Belt B, Miyashiro K, McIntosh T, Eberwine J (2004) Single neurons as experimental systems in molecular biology. Prog Neurobiol 72(2):129–142
77. Chiang MK, Melton DA (2003) Single-cell transcript analysis of pancreas development. Dev Cell 4(3):383–393
78. Kralj JG, Player A, Sedrick H, Munson MS, Petersen D, Forry SP, Meltzer P, Kawasaki E, Locascio LE (2009) T7-based linear amplification of low concentration mRNA samples

using beads and microfluidics for global gene expression measurements. Lab Chip 9(7):917–924
79. Gong Y, Ogunniyi AO, Love JC (2010) Massively parallel detection of gene expression in single cells using subnanolitre wells. Lab Chip: Epub ahead of print
80. Tang F, Barbacioru C, Wang Y, Nordman E, Lee C, Xu N, Wang X, Bodeau J, Tuch BB, Siddiqui A, Lao K, Surani MA (2009) mRNA-Seq whole-transcriptome analysis of a single cell. Nat Methods 6(5):377–382
81. Filiatrault MJ, Stodghill PV, Bronstein PA, Moll S, Lindeberg M, Grills G, Schweitzer P, Wang W, Schroth GP, Luo S, Khrebtukova I, Yang Y, Thannhauser T, Butcher BG, Cartinhour S, Schneider DJ (2010) Transcriptome analysis of *Pseudomonas syringae* identifies new genes, noncoding RNAs, and antisense activity. J Bacteriol 192(9):2359–2372
82. Isaacs FJ, Blake WJ, Collins JJ (2005) Molecular biology. Signal processing in single cells. Science 307(5717):1886–1888
83. Lidstrom ME, Meldrum DR (2003) Life-on-a-chip. Nat Rev Microbiol 1(2):158–164
84. Le TT, Harlepp S, Guet CC, Dittmar K, Emonet T, Pan T, Cluzel P (2005) Real-time RNA profiling within a single bacterium. Proc Natl Acad Sci USA 102(26):9160–9164
85. Shav-Tal Y, Darzacq X, Shenoy SM, Fusco D, Janicki SM, Spector DL, Singer RH (2004) Dynamics of single mRNPs in nuclei of living cells. Science 304(5678):1797–1800
86. Paulsson J (2004) Summing up the noise in gene networks. Nature 427(6973):415–418
87. Bertrand E, Chartrand P, Schaefer M, Shenoy SM, Singer RH, Long RM (1998) Localization of ASH1 mRNA particles in living yeast. Mol Cell 2(4):437–445
88. Golding I, Cox EC (2004) RNA dynamics in live *Escherichia coli* cells. Proc Natl Acad Sci USA 101(31):11310–11315
89. Shen Y, Tolic N, Masselon C, Pasa-Tolic L, Camp DG 2nd, Hixson KK, Zhao R, Anderson GA, Smith RD (2004) Ultrasensitive proteomics using high-efficiency on-line micro-SPE-nanoLC-nanoESI MS and MS/MS. Anal Chem 76(1):144–154
90. Aebersold R, Mann M (2003) Mass spectrometry-based proteomics. Nature 422(6928):198–207
91. Hu S, Le Z, Krylov S, Dovichi NJ (2003) Cell cycle-dependent protein fingerprint from a single cancer cell: image cytometry coupled with single-cell capillary sieving electrophoresis. Anal Chem 75(14):3495–3501
92. Hu S, Le Z, Newitt R, Aebersold R, Kraly JR, Jones M, Dovichi NJ (2003) Identification of proteins in single-cell capillary electrophoresis fingerprints based on comigration with standard proteins. Anal Chem 75(14):3502–3505
93. Gao J, Yin XF, Fang ZL (2004) Integration of single cell injection, cell lysis, separation and detection of intracellular constituents on a microfluidic chip. Lab Chip 4(1):47–52
94. Han F, Wang Y, Sims CE, Bachman M, Chang R, Li GP, Allbritton NL (2003) Fast electrical lysis of cells for capillary electrophoresis. Anal Chem 75(15):3688–3696
95. Zhang Z, Krylov S, Arriaga EA, Polakowski R, Dovichi NJ (2000) One-dimensional protein analysis of an HT29 human colon adenocarcinoma cell. Anal Chem 72(2):318–322
96. Hellmich W, Pelargus C, Leffhalm K, Ros A, Anselmetti D (2005) Single cell manipulation, analytics, and label-free protein detection in microfluidic devices for systems nanobiology. Electrophoresis 26(19):3689–3696
97. Malek A, Khaledi MG (1999) Expression and analysis of green fluorescent proteins in human embryonic kidney cells by capillary electrophoresis. Anal Biochem 268(2):262–269
98. Turner EH, Lauterbach K, Pugsley HR, Palmer VR, Dovichi NJ (2007) Detection of green fluorescent protein in a single bacterium by capillary electrophoresis with laser-induced fluorescence. Anal Chem 79(2):778–781
99. Zhang H, Jin W (2004) Determination of different forms of human interferon-gamma in single natural killer cells by capillary electrophoresis with on-capillary immunoreaction and laser-induced fluorescence detection. Electrophoresis 25(7–8):1090–1095

100. Michels DA, Hu S, Schoenherr RM, Eggertson MJ, Dovichi NJ (2002) Fully automated two-dimensional capillary electrophoresis for high sensitivity protein analysis. Mol Cell Proteomics 1(1):69–74
101. Zhang H, Jin W (2006) Single-cell analysis by intracellular immuno-reaction and capillary electrophoresis with laser-induced fluorescence detection. J Chromatogr A 1104(1–2): 346–351
102. Sun X, Jin W (2003) Catalysis-electrochemical determination of zeptomole enzyme and its application for single-cell analysis. Anal Chem 75(22):6050–6055
103. Li L, Garden RW, Romanova EV, Sweedler JV (1999) In situ sequencing of peptides from biological tissues and single cells using MALDI-PSD/CID analysis. Anal Chem 71(24):5451–5458
104. Rubakhin SS, Greenough WT, Sweedler JV (2003) Spatial profiling with MALDI MS: distribution of neuropeptides within single neurons. Anal Chem 75(20):5374–5380
105. Rubakhin SS, Sweedler JV (2008) Quantitative measurements of cell-cell signaling peptides with single-cell MALDI MS. Anal Chem 80(18):7128–7136
106. Rubakhin SS, Churchill JD, Greenough WT, Sweedler JV (2006) Profiling signaling peptides in single mammalian cells using mass spectrometry. Anal Chem 78(20):7267–7272
107. Rubakhin SS, Sweedler JV (2007) Characterizing peptides in individual mammalian cells using mass spectrometry. Nat Protoc 2(8):1987–1997
108. Jehmlich N, Hubschmann T, Salazar MG, Volker U, Benndorf D, Muller S, Bergen Mv, Schmidt F (2010) Advanced tool for characterization of microbial cultures by combining cytomics and proteomics. Appl Microbiol Biotechnol in press.
109. Hofstadler SA, Severs JC, Smith RD, Swanek FD, Ewing AG (1996) Analysis of single cells with capillary electrophoresis electrospray ionization Fourier transform ion cyclotron resonance mass spectrometry. Rapid Commun Mass Spectrom 10(8):919–922
110. Cao P, Moini M (1999) Separation and detection of the alpha- and beta-chains of hemoglobin of a single intact red blood cells using capillary electrophoresis/electrospray ionization time-of-flight mass spectrometry. J Am Soc Mass Spectrom 10(2):184–186
111. Moini M, Demars SM, Huang H (2002) Analysis of carbonic anhydrase in human red Blood cells using capillary electrophoresis/electrospray ionization-mass spectrometry. Anal Chem 74(15):3772–3776
112. Belov ME, Gorshkov MV, Udseth HR, Anderson GA, Smith RD (2000) Zeptomole-sensitivity electrospray ionization-Fourier transform ion cyclotron resonance mass spectrometry of proteins. Anal Chem 72(10):2271–2279
113. Shen Y, Tolic N, Masselon C, Pasa-Tolic L, Camp DG 2nd, Lipton MS, Anderson GA, Smith RD (2004) Nanoscale proteomics. Anal Bioanal Chem 378(4):1037–1045
114. Newman JR, Ghaemmaghami S, Ihmels J, Breslow DK, Noble M, DeRisi JL, Weissman JS (2006) Single-cell proteomic analysis of *S. cerevisiae* reveals the architecture of biological noise. Nature 441(7095):840–846
115. Cohen AA, Geva-Zatorsky N, Eden E, Frenkel-Morgenstern M, Issaeva I, Sigal A, Milo R, Cohen-Saidon C, Liron Y, Kam Z, Cohen L, Danon T, Perzov N, Alon U (2008) Dynamic proteomics of individual cancer cells in response to a drug. Science 322(5907):1511–1516
116. Yin H, Killeen K, Brennen R, Sobek D, Werlich M, van de Goor T (2005) Microfluidic chip for peptide analysis with an integrated HPLC column, sample enrichment column, and nanoelectrospray tip. Anal Chem 77(2):527–533
117. Fortier MH, Bonneil E, Goodley P, Thibault P (2005) Integrated microfluidic device for mass spectrometry-based proteomics and its application to biomarker discovery programs. Anal Chem 77(6):1631–1640
118. Jo K, Heien ML, Thompson LB, Zhong M, Nuzzo RG, Sweedler JV (2007) Mass spectrometric imaging of peptide release from neuronal cells within microfluidic devices. Lab Chip 7(11):1454–1460
119. Ryley J, Pereira-Smith OM (2006) Microfluidics device for single cell gene expression analysis in *Saccharomyces cerevisiae*. Yeast 23(14–15):1065–1073

120. Yu J, Xiao J, Ren XJ, Lao KQ, Xie XS (2006) Probing gene expression in live cells, one protein molecule at a time. Science 311(5767):1600–1603
121. Gefen O, Gabay C, Mumcuoglu M, Engel G, Balaban NQ (2008) Single-cell protein induction dynamics reveals a period of vulnerability to antibiotics in persister bacteria. Proc Natl Acad Sci USA 105(16):6145–6149
122. Kortmann H, Kurth F, Blank LM, Dittrich P, Schmid A (2009) Towards real time analysis of protein secretion from single cells. Lab Chip 9(21):3047–3049
123. Amantonico A, Oh JY, Sobek J, Heinemann M, Zenobi R (2008) Mass spectrometric method for analyzing metabolites in yeast with single cell sensitivity. Angew Chem Int Ed Engl 47(29):5382–5385
124. Schrumpf B, Eggeling L, Sahm H (1992) Isolation and prominent characteristics of an L-lysine hyperproducing strain of *Corynebacterium glutamicum*. Appl Microbiol Biotechnol 37(5):566–571
125. Blank LM, Kuepfer L, Sauer U (2005) Large-scale ^{13}C-flux analysis reveals mechanistic principles of metabolic network robustness to null mutations in yeast. Genome Biol 6(6):R49
126. Krylov SN, Zhang Z, Chan NW, Arriaga E, Palcic MM, Dovichi NJ (1999) Correlating cell cycle with metabolism in single cells: combination of image and metabolic cytometry. Cytometry 37(1):14–20
127. Whitmore CD, Hindsgaul O, Palcic MM, Schnaar RL, Dovichi NJ (2007) Metabolic cytometry. Glycosphingolipid metabolism in single cells. Anal Chem 79(14):5139–5142
128. Zhang H, Jin W (2004) Analysis of amino acids in individual human erythrocytes by capillary electrophoresis with electroporation for intracellular derivatization and laser-induced fluorescence detection. Electrophoresis 25(3):480–486
129. Allen PB, Doepker BR, Chiu DT (2009) High-throughput capillary-electrophoresis analysis of the contents of a single mitochondria. Anal Chem 81(10):3784–3791
130. Wu H, Wheeler A, Zare RN (2004) Chemical cytometry on a picoliter-scale integrated microfluidic chip. Proc Natl Acad Sci USA 101(35):12809–12813
131. Lapainis T, Rubakhin SS, Sweedler JV (2009) Capillary electrophoresis with electrospray ionization mass spectrometric detection for single-cell metabolomics. Anal Chem 81(14):5858–5864
132. Mizuno H, Tsuyama N, Date S, Harada T, Masujima T (2008) Live single-cell metabolomics of tryptophan and histidine metabolites in a rat basophil leukemia cell. Anal Sci 24(12):1525–1527
133. Amantonico A, Urban PL, Oh JY, Zenobi R (2009) Interfacing microfluidics and laser desorption/Ionization mass spectrometry by continuous deposition for application in single cell analysis. Chimia (Aarau) 63(4):185–188
134. Deuschle K, Chaudhuri B, Okumoto S, Lager I, Lalonde S, Frommer WB (2006) Rapid metabolism of glucose detected with FRET glucose nanosensors in epidermal cells and intact roots of *Arabidopsis* RNA-silencing mutants. Plant Cell 18(9):2314–2325
135. Niittylae T, Chaudhuri B, Sauer U, Frommer WB (2009) Comparison of quantitative metabolite imaging tools and carbon-13 techniques for fluxomics. Methods Mol Biol 553:355–372
136. Cai L, Friedman N, Xie XS (2006) Stochastic protein expression in individual cells at the single molecule level. Nature 440(7082):358–362
137. Cai X, Klauke N, Glidle A, Cobbold P, Smith GL, Cooper JM (2002) Ultra-low-volume, real-time measurements of lactate from the single heart cell using microsystems technology. Anal Chem 74(4):908–914
138. Cheng W, Klauke N, Sedgwick H, Smith GL, Cooper JM (2006) Metabolic monitoring of the electrically stimulated single heart cell within a microfluidic platform. Lab Chip 6(11):1424–1431
139. Molter TW, Holl MR, Dragavon JM, McQuaide SC, Anderson JB, Young AC, Burgess LW, Lidstrom ME, Meldrum DR (2008) A new approach for measuring single-cell oxygen consumption rates. IEEE Trans Autom Sci Eng 5(1):32–42

140. Dragavon J, Molter T, Young C, Strovas T, McQuaide S, Holl M, Zhang M, Cookson B, Jen A, Lidstrom M, Meldrum D, Burgess L (2008) A cellular isolation system for real-time single-cell oxygen consumption monitoring. J R Soc Interface 5 Suppl 2:S151–S159
141. Molter TW, McQuaide SC, Suchorolski MT, Strovas TJ, Burgess LW, Meldrum DR, Lidstrom ME (2009) A microwell array device capable of measuring single-cell oxygen consumption rates. Sens Actuators B Chem 135(2):678–686
142. Diaz-Ruiz R, Uribe-Carvajal S, Devin A, Rigoulet M (2009) Tumor cell energy metabolism and its common features with yeast metabolism. Biochim Biophys Acta 1796(2):252–265
143. Chao TC, Ros A (2008) Microfluidic single-cell analysis of intracellular compounds. J R Soc Interface 5:S139–S150
144. Andersson H, van den Berg A (2004) Microtechnologies and nanotechnologies for single-cell analysis. Curr Opin Biotechnol 15(1):44–49
145. Sims CE, Allbritton NL (2007) Analysis of single mammalian cells on-chip. Lab Chip 7(4):423–440
146. Huang WH, Ai F, Wang ZL, Cheng JK (2008) Recent advances in single-cell analysis using capillary electrophoresis and microfluidic devices. J Chromatogr B Analyt Technol Biomed Life Sci 866(1–2):104–122
147. Borland LM, Kottegoda S, Phillips KS, Allbritton NL (2008) Chemical analysis of single cells. Annu Rev Anal Chem 1:191–227
148. Roman GT, Chen YL, Viberg P, Culbertson AH, Culbertson CT (2007) Single-cell manipulation and analysis using microfluidic devices. Anal Bioanal Chem 387(1):9–12
149. Schmid A, Blank LM (2010) Systems biology: hypothesis-driven omics integration. Nat Chem Biol 6(7):485–487

Adv Biochem Engin/Biotechnol (2011) 124: 123–150
DOI: 10.1007/10_2010_95

Published Online: 11 November 2010

Cultivation-independent Assessment of Bacterial Viability

Frederik Hammes, Michael Berney and Thomas Egli

Abstract Cultivation-independent assessment of bacterial viability is essential when (1) results are required fast and at high throughput, and/or (2) when the specific target or mode-of-action of a certain bactericidal process is of interest, and/or (3) when the organisms under investigation are regarded as "uncultivable". However, aside from cultivation, there exists no "silver bullet" method that demonstrates with absolute certainty whether an organism is alive or dead, and all currently available methods are prone to produce varying results with different organisms and in different environments. Here we discuss the fundamental concept of viability in bacteria, with specific focus on the main aspects that define it. It is argued that the presence of intact and functional nucleic acids, as well as an intact and polarized cytoplasmic membrane are essential components of cellular viability, while numerous other parameters and processes that are linked to viability are explored. Different methods/approaches are discussed with particular emphasis on the advantages and disadvantages of each approach, the applicability of the methods toward environmental samples, and the underlying link between the various viability parameters.

Keywords Activity · Bacteria · Flow cytometry · Membrane potential · Viability

---

F. Hammes (✉) and T. Egli
Eawag, Swiss Federal Institute of Aquatic Science and Technology,
8600 Dübendorf, Switzerland
e-mail: frederik.hammes@eawag.ch

T. Egli
e-mail: thomas.egli@eawag.ch

M. Berney
Department of Microbiology and Immunology, Otago School of Medical Sciences,
University of Otago, P.O. Box 56, Dunedin, New Zealand
e-mail: michael.berney@otago.ac.nz

Contents

1 Introduction

Knowing what is alive and what is dead, as well as what transpires in between those two extremities, is one of the main challenges that microbiologists are facing across all fields of application on a daily basis. The information arising from this knowledge affects people in their daily lives. For example, the production of wine, beer, cheese and yogurt requires well-defined viable microbial starting cultures as well as quality control during the production process [44]. Similarly, the productivity of industrial fermentations that utilise microbial cultures to synthesize metabolites and recombinant proteins can be controlled and optimised through real-time monitoring and increased understanding of the viability and activity of the vector organism [41, 70]. Drinking water and wastewater treatment systems often utilise disinfection processes as critical steps to safeguard the public against microbiological diseases. The dosages of disinfectants and the constant monitoring of the efficacy of such processes, are directly dependent on accurate viability assessment [23, 53, 72]. In medical-related research, a well-known application of viability assessment is testing the efficacy and mode of action of current and new antibiotics [42, 51]. Furthermore, viability assessment is used in research laboratories worldwide to understand the fundamental processes that drive growth, survival and die-off of bacteria in both carefully controlled laboratory environments [4] as well as in complex natural ecosystems [63].

While accurate viability assessment of bacteria is clearly important, it is far from simple. First and foremost, it boils down to the essential question of what exactly defines "life" and "death" for bacteria? This becomes a near-philosophical discussion, initiated already half a decade ago by Postgate [56] and aptly summarised by Roszak and Colwell [60], in which opinions remain highly subjective. Secondly, viability assessment requires a fundamental understanding of the various parameters that constitute and define life or death for a bacterium, relative to both the individual organism and the environment in which it is studied. Moreover, it requires tools that can determine these viability parameters in an accurate and unambiguous manner, also distinguishing between parameters that change immediately upon cell death, and those that are affected in a time-dependent manner. As will be discussed below, these issues alone already pose daunting challenges to researchers and practitioners. Thirdly, viability assessment is complicated by the fact that bacteria represent a broad and complex group of organisms [66]. Based on 16S rRNA sequences, there are in excess of 50 recognized bacterial phyla [58, 62], which in layman's terms mean groups of bacteria as genetically distinct from each other as a snail is from a human. There are in these phyla an estimated 1–10 million bacterial species, and these organisms exhibit immense heterogeneity on multiple levels including physical properties (e.g., size, shape and composition), nutritional behaviour (e.g., oligotrophy vs. copiotrophy) and physiological states (e.g., dormancy or exponential growth). For a simple example, some organisms have Gram-positive cell walls, while others have Gram-negative cell walls, S-layers, sheaths, etc. The cell wall is the primary barrier between the cell and the environment, and the composition thereof is known to affect the action, functionality and interpretation of some commonly used viability stains [5, 66]. Some bacteria are able to thrive in a high cell density fast growing environment (e.g., fed-batch bioreactors) while others can persist in low cell concentrations under near-starvation conditions (e.g., groundwater). Different growth conditions are likely to affect the way in which bacteria are perceived in terms of viability and activity (e.g., RNA content or growth rates). Moreover, bacterial heterogeneity is not limited to broad groups, but is known to occur even within a single species growing in the same environment [13, 47], further highlighting the need for meaningful analysis on single-cell level.

Technological advances during the last 30 years, specifically the development of powerful epi-fluorescence microscopes and the accompanying methodologies of fluorescent staining, have largely facilitated the rise of fast, cultivation-independent microbial methods. Moreover, the focus has shifted from analysing bulk sample parameters to analysing individual organisms on single-cell level [13]. In fact, improvements in laser technology and microfluidics have pushed single-cell analysis to the point where basic bench top flow cytometers are common instruments in research laboratories [22], and where inline/online instrumentation is a reality rather than a vision [3, 18]. These practical advances have facilitated fundamental shifts in the perception and understanding of microbial diversity and heterogeneity, both within pure cultures and in natural environments [13, 47].

This paper approaches cultivation-independent viability assessment from the perspective of "*what is life or death for a bacterium*" and how various viability parameters relate to this. The purpose is to demonstrate the subjective nature of the topic, to provide insights in how this topic can be approached in research and/or routine monitoring, as well as to highlight the practical value gained from such information. Specific focus is given to high throughput single-cell methods, typically requiring microscopy or flow cytometry for analysis. The first section covers the cultivation paradox, highlighting the need for cultivation-independent methods for viability assessment, while the second part explores the fundamental aspects of bacterial viability. In the third section, different bacterial processes related to viability are discussed in the context of tools used for the analysis of these processes. The take home message is that the proverbial "silver bullet" for viability assessment remains (and will probably remain) illusive, but an intelligent combination of available methods can be (and have already been) used to obtain interesting and new information on diverse aspects of bacterial viability.

For additional reading, five recent review papers address the current state-of-the-science with broader focus on flow cytometric applications in general [15, 16, 22, 46, 67], while the history of cultivation-independent viability research is outlined in the thorough review papers of Roszak and Colwell [60], McFeters et al. [45], Kell et al. [34], Nebe-von Caron et al. [49] and Joux and Lebaron [31].

2 Viability Versus Cultivability

2.1 Cultivability: The Gold Standard

While the present paper deals specifically with cultivation-independent viability analysis, the basic fact is that cultivation (including conventional plating, most probable number (MPN), and direct viable counts (DVC) remains the single most solid evidence of viability [34], if the appropriate growth medium and cultivation conditions for the organism are known [60]. All bacteria are in theory cultivable, for the simple reason that they have grown/divided in order to exist. Moreover, at least one school of thought on the topic of viability argues that an organism has to be cultivable in order to be classified as viable [56, 60], and several disciplines (e.g., drinking water analysis) still rely heavily on cultivation as primary method for microbial analysis. However, numerous strong arguments exist in favour of looking beyond cultivation alone when describing viability, and these are explored below.

2.2 Time and Throughput

Cultivation can be a time-consuming analysis method, with growth to detectable colony formation taking from 1 day to well over 1 month, depending on the organisms and the growth media. Notably, some advanced cultivation methods (e.g., micro-colony counting and DVC) improve the time problem to some extent [2].

Nonetheless, there are numerous cases, particularly in industrial and hygienic applications, where a long time lapse would not suffice the purpose of the analysis. For example, if a microbiological problem is detected in a drinking water disinfection step only 3 days after the actual event, any corrective action will only occur after the water has already reached the consumer. Similar examples can be found in food and beverage industries. Indeed, the aim for microbiological analysis should be to achieve accurate measurements in a matter of minutes, rather than hours or days. Moreover, cultivation is a labour intensive method that does not allow high throughput detection similar to automated microscopy, solid-phase cytometry or flow cytometry.

2.3 Dealing with So-called "Uncultivable" Bacteria

One of the strongest arguments in favour of cultivation-independent analysis is the analysis of indigenous environmental bacteria. Firstly, for the vast majority of environmental bacteria, appropriate cultivation methods are simply not known [58, 69], thus nullifying the use of conventional cultivation-based approaches. Secondly, it is highly unlikely that all organisms in the same environmental sample would share the same synthetic cultivation medium, rendering this approach inappropriate for analysis of a diverse indigenous microbial community. Notably, some recent studies have addressed this problem by cultivating bacteria in sterile media prepared from the same aqueous environments from which the targeted organisms originated (e.g., [69]). Thirdly, several reasons such as extremely slow growth of specific organisms might render cultivation impractical to detect [49]. Organisms that do not grow readily on conventional growth media are broadly (and most probably erroneously) referred to as "uncultivable/unculturable" bacteria. In the case of environmental samples, it is clear that only cultivation-independent analysis can provide the much-needed information when viability is assessed, which makes it the most appropriate approach for analysis of samples from e.g., wastewater, drinking water and natural surface waters [6, 63].

2.4 The Viable-But-Not-Cultivable (VBNC) Paradigm

The possibility of VBNC bacteria is often highlighted as a key reason for doing cultivation-independent viability analysis [28]. In short: the argument is that normally cultivable bacteria under certain conditions become uncultivable on media that they previously grew on, while maintaining other measurable viability signs [34, 52]. However, it would not be over-critical to state that the VBNC term is used very liberally in peer-reviewed literature, often with scant regard for the basic concepts of viability, which it apparently describes. The most common VBNC example is the use of temperature shifts to regulate cultivability of *Vibrio vulnificus*, with cold-stored cells losing their ability to grow on standard media, and regaining it through a temperature upshift [52]. Another example was shown by Berney et al. [4] with *E. coli* exposed to low levels of solar radiation stress, no longer growing on the standard cultivation media, but still appearing viable with cultivation-independent parameters. At first sight, these two examples seem to underpin the VBNC concept. However, in the latter case of the stressed *E. coli* cells, it was demonstrated that a considerable percentage of the "uncultivable" cells were capable of growing anaerobically on medium that was specifically supplemented with sodium pyruvate (which scavenges reactive oxygen species) [4], showing that the cells were indeed cultivable if the appropriate medium is used. Similarly, Bogosian and Bourneuf [7] described a control experiment that showed that the cold-stored uncultivable *V. vulnificus* cells were probably injured cells that were merely sensitive to the high concentrations of hydrogen peroxide found in rich cultivation media. In fact, the brilliant critical review by Kell et al. [34], later mirrored by Bogosian and Bourneuf [7], needs to be illuminated in the VBNC context. This has to serve as a blueprint for VBNC advocates for the experimental standard to which this concept should be held, and the importance of careful use of this terminology. The somewhat critical opinion of VBNC data expressed herein is not meant to undermine the efforts of understanding different physiological states in bacteria, but rather to emphasize the care that should be taken with viability analysis tools, and the expression of data generated with these methods.

2.5 A Description of Specific Injuries and Slow Death

There is another clear advantage of cultivation-independent methods over cultivation-based viability methods. Cultivation provides only a binary (presence/absence) result, i.e. the organisms are either detected as cultivable or not, a result which is then usually interpreted as meaning alive or dead. But this type of result limits the researcher in his/her understanding of the process that brought about the result. The fundamental aspects of cell death, such as the sequence of damage or type of injury caused by a certain bactericidal agent (e.g., [8]), or the specific targets of different bactericidal agents and the kinetics of damage to these targets (e.g., [51]), all requires specific tools that can explore individual cellular processes and mechanisms in more detail than a simplistic yes/no answer. The fact that cultivation-independent viability analysis targets a variety of individual cellular processes renders this approach meaningful and suitable to address such complex questions.

3 How Dead is Dead?

In the absence of cultivation-based methods, the determination of viability becomes a difficult and contentious topic. Argumentatively, three main cellular components are prerequisites for life, namely: (1) the presence of functional

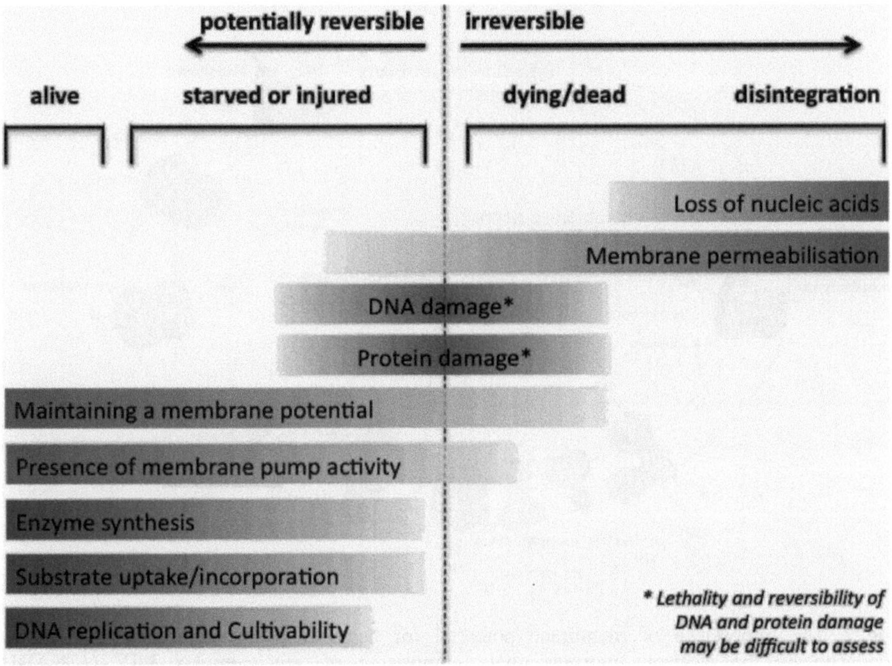

Fig. 1 The four main stages of bacterial cell viability and the cellular processes that relate to these stages. The order and degree of the affected processes, as well as the reversibility of the injury, depends on the organism, the cause of injury/death, and the environment (adapted from [16, 49])

nucleic acids, allowing transcription/translation and DNA replication, (2) the presence of minimum cellular energy, allowing basic functioning of cellular processes, and (3) the presence of an intact and functional cellular membrane, maintaining the unique intracellular environment (Fig. 1). These three components are inherently linked to one another as well as with other cellular components and viability parameters, and without them, a cell cannot be considered as viable.

For all bacteria, one of the core processes essential for life is the maintenance of a membrane potential. The membrane potential is the voltage difference between the interior and exterior of a cell (usually between 70 and 200 mV in bacteria) and in microorganisms this is typically generated by the electron transport chain (proton translocation) or enzymes like oxaloacetate decarboxylase [37] that can translocate sodium ions (Fig. 2). The membrane potential ($\Delta\varphi$) and the pH gradient ($Z\Delta pH$) produce the proton motive force (also called the electrochemical gradient), which drives ATP synthesis via the F_1F_O ATP synthase in most bacteria [17]. Membrane potential is used for (1) generating energy (ATP synthase), (2) driving active transport of molecules across the membrane (such as the active efflux/uptake of potassium ions), (3) enabling motility (flagella), and (4) keeping the cytoplasm from equilibrating with the environment (like maintaining a specific intracellular pH) (Fig. 2). Technically, a bacterial cell that cannot maintain a

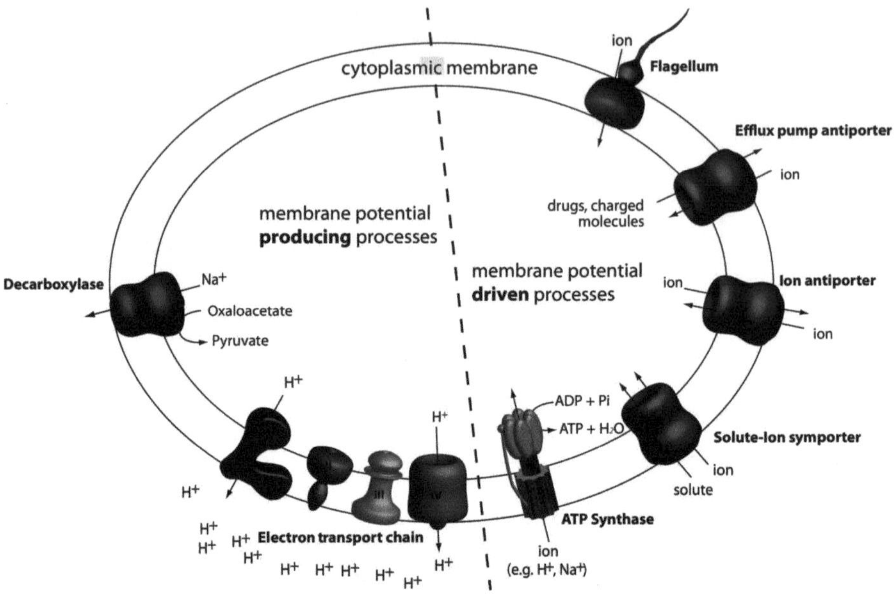

Fig. 2 The importance of membrane potential for bacterial cells. Membrane potential is produced by the electron transport chain comprising of, e.g. complex I–IV (*I* NADH dehydrogenase, *II* succinate dehydrogenase, *III* cytochrome bc_1 complex, *IV* terminal oxidase, e.g. cytochrome *c* oxidase) and/or sodium-pumping decarboxylases (e.g., oxaloacetate decarboxylase). Membrane potential powers processes like the ATP synthase, solute-ion symporter, e.g. (porline/Na^+), ion antiproters (e.g., H^+/K^+), efflux pumps (e.g., ethidium) and motility

membrane potential can be considered dead, since this will lead to an inhibition of selective exchange of molecules across the cytoplasmic membrane like ions, metabolites, amino acids or carbon and energy sources. For example, the maintenance of a high intracellular concentration of potassium is essential for protein and nucleic acid synthesis and turgor pressure. The difference between intracellular and extracellular potassium concentration can be up to 100-fold (5 M intracellular, 0.05 M extracellular) as for the bacterium *Halobacterium salinarium* [43]. Likewise, sodium ion homeostasis is essential for numerous sodium/solute symporters present in bacterial membranes [32]. Generally speaking, without a membrane potential, cells cannot maintain an intracellular environment that enables the functioning of life-supporting metabolic processes and the cells become more vulnerable to their environment. This additionally means that most other cellular functions that are measured with viability indicators (e.g., efflux pump activity, substrate uptake, ATP synthesis) are dependent or linked to a membrane potential. Hence, measuring the membrane potential, or the dissipation thereof, is in theory a good discriminator between living and dead cells.

However, a cornerstone to the above argument in favour of membrane potential as a main viability indicator is the principle that a bacterium should be able to "*maintain*" its membrane potential: a bacterium that experiences a lethal stress does not necessarily disintegrate immediately, nor does it lose its membrane potential on the spot. In most cases, death occurs with a time-dependent shutdown of cellular processes (in minutes or hours) that is dependent on the type of lethal agent (e.g., antibiotics, H_2O_2, chlorine, UV light or starvation), the extent of the cellular damage, and the ability of the cell to recover form injury (Fig. 1). For example, cells of *Salmonella typhimurium* that have been exposed to UV-A light for 2 h, retained a membrane potential and the ability to actively transport glucose immediately after irradiation but the majority of the cells lost these abilities after 24 h dark storage [8]. These same authors demonstrated in a subsequent study that a major cause of cell death is irreversible damage to cellular proteins [9]. This is a clear example of lethal injury requiring a certain timeframe to manifest in detectable viability parameters, highlighting the dangers of rash interpretations of viability staining.

4 Specific Processes that can be Measured with Viability Indicators

The following section covers a variety of cellular processes, functions and parameters that are linked to bacterial viability, and for which in many cases well-developed methodology is already available (Fig. 3). It deals mostly with

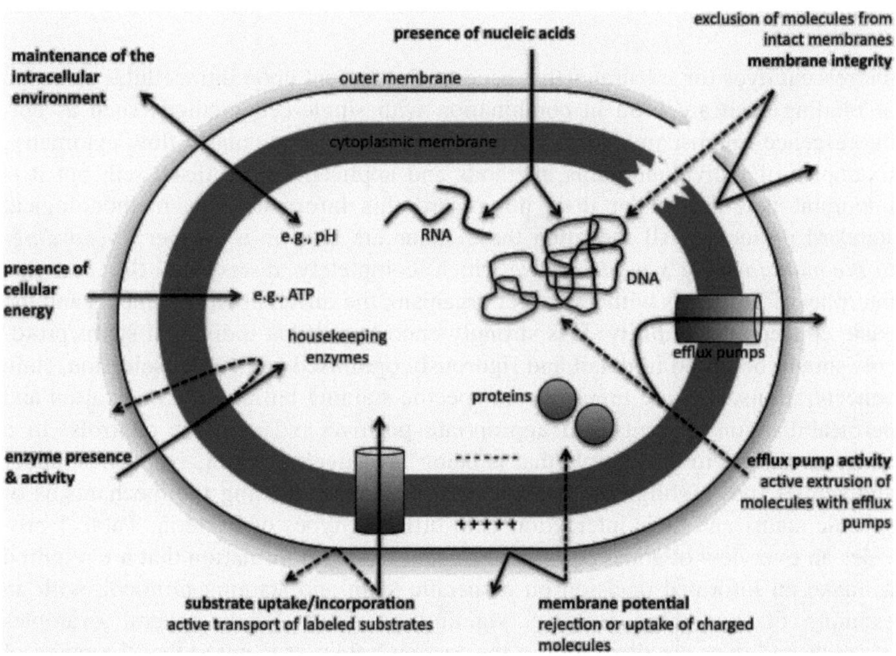

Fig. 3 The concept of evaluating various cellular processes in order to assess the viability state of the organism (adapted from [4, 16, 31])

Table 1 Essential criteria and required information for the selection and optimisation of a specific stain

Parameter	Information	Example
Stain selection	Stain	Propidium iodide
	Purported mechanism	Exclusion based on size and charge
	Cellular binding target	Nucleic acids
	Binding mechanism	Intercalates with DNA
	Non-specific binding	Not known
	Species specificity	Non-specific
	Instrumentation	EFM, FCM
Staining conditions	Concentration	2–6 μM for $<10^6$ cells/mL
	Temperature	>20 °C
	Staining time (min/max)	15–30 min
	Salinity, pH, etc.	Not known
Additives required	Permeabilisation	EDTA (5 mM)
	Potential for false results	Increased positives with EDTA use
	Stain combinations	SYTO 9, SYBR Green I
Controls	Positive control	Cell damage with lysis buffer or heat
	Negative control	Freshly grown pure culture
Exceptions	Known false positives	Possible (e.g. [65])
Alternatives	Dyes with similar actions	SYTOX green, Ethidium homodimer2

The example of membrane integrity staining with propidium iodide is provided, based on experience in the authors' group [5, 6]

fluorescent dyes (or substrates that become fluorescent upon intracellular cleavage or binding) that are used in combination with single-cell methods such as epifluorescence microscopy, solid-phase cytometry and particularly flow cytometry. Examples of individual stains, methods and applications are discussed, but it is important that the reader does not regard this information as methodological standard protocols. All too often these stains are used in a manner "*according-to-the-manufacturer's-guidelines*", which completely disrespects the complex interplay of the stains with the target organism, the different environments and the basic concepts of viability. It is strongly encouraged that individual stains/protocols should be tested in detail and rigorously optimised (e.g., stain selection, stain concentrations, staining times, use of specific staining buffers, use of fixation and permeabilisation reagents and appropriate positive and negative controls) in a manner relevant to the sample that is being investigated. In this respect, Sträuber and Müller [66] highlighted the importance of understanding the mechanisms of specific stains and their interaction with different types of bacteria. Table 1 provides an overview of some of the criteria and related information that are required to make an informed decision on a specific stain and staining protocol, with an example of membrane integrity staining provided. While several examples of stains and dyes are discussed in the section below, it is noted that the range of available products is significantly broader and expanding on a regular basis. A good tabular overview of viability stains can be found in Tracy et al. [67].

The different processes are discussed as individual entities for the sake of clarity, but it is emphasized that these separate processes are often linked directly to each other, forming part of a broader viability concept or continuum (Fig. 1) [49]. Therefore, these processes (and the related methodologies) should preferably not be viewed or used in isolation. Moreover, it is important to realise that the parameters measured in typical viability assessments are usually not measuring viability as such, but rather provide information about a specific cellular function that can be related to viability (Fig. 3). Hence, if the mode of action of a bactericidal agent is known (e.g., oxidative membrane damage by chlorine dioxide), specific viability indicators can be used to assess the mode of action of interest. However, if a biocidal agent has multiple target sites (e.g., UV-A light), or if a completely unknown sample is analysed (e.g., natural surface water), viability stains are best used in concert (e.g., [4, 28, 40, 51, 70]). This enables the researcher to derive a meaningful interpretation of the sample in question.

4.1 Presence of (Intact) Nucleic Acids

4.1.1 Principle

Without intact and functional nucleic acids, a cell will not be able to replicate or produce any proteins and thus perform even the most basic cellular functions including repair, stress response and cell division. Cells without nucleic acids are described in literature as ghost cells [31], and are regarded as irreversibly dead (Fig. 1).

4.1.2 Methodology

Well-known total cell count stains such as SYTO 9, SYBR Green I and II, PicoGreen and Acridine Orange can be used to stain the nucleic acids of bacteria [23, 38]. From a practical perspective, it is important to ensure that the outer membrane is properly permeabilised (e.g., by adding EDTA) during nucleic acid staining, since it was shown previously that some Gram-negative bacteria partially reject such stains (Fig. 4) [5]. Schumann et al. [63] makes the relevant point that partially disintegrated dead cells without a chromosome would go undetected with any nucleic acid targeting stains, and this was elucidated by Phe et al. [54], who showed that propidium iodide cannot bind to cellular nucleic acids that were damaged extensively by progressive chlorination, leaving such cells undetected. Argumentatively, cells that are damaged to an extent that nucleic acids can no longer be detected should perhaps not even be recognised as cells, but rather as organic particles. As an alternative, a recent publication by Saint-Ruf et al. [61] proposed the use of highly fluorescent hydrazides that do not bind to nucleic acids but rather to carbonyl groups on intracellular proteins, for the detection of dead cells where nucleic acids might already be absent or damaged to an extent that does not allow binding of conventional nucleic acid stains. However, while the absence of nucleic

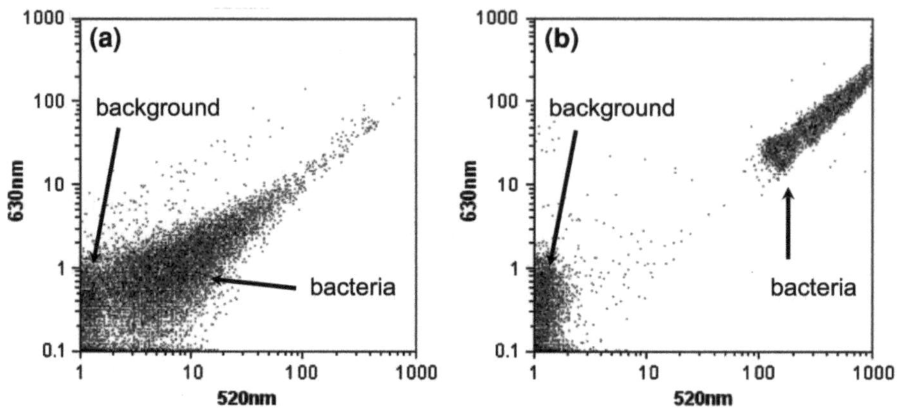

Fig. 4 The potential impact of pre-treatment in staining protocols: nucleic acid labelling of stationary phase Gram-negative *E. coli* cells with SYBR Green I without (**a**) and with (**b**) a pre-treatment step using a commercially available lysis buffer. Staining was done for 15 min at 25 °C; analysis was done with a Cyflow Space instrument (Partec)

acids is a certain "death" indicator, the mere presence of nucleic acids is not a measurement of viability. For example, none of the general stains that are mentioned above are sensitive enough to detect the specific nucleic acid damage caused by UV-C disinfection, commonly used in drinking water treatment. The inactivation of bacteria by far-ultraviolet (190–300 nm) radiation results from the absorption of the radiation by the DNA, causing the formation of thymine dimers that distort the conformation of the DNA double helix and interfere with cell replication and transcription [1, 25]. A recent paper describes the use of a monoclonal antibody against cyclobutyl thymine dimers (anti-TDmAb) that results in dimer specific fluorescence [1]. This can be used on a single-cell level and has been demonstrated to detect UV damage in protozoan parasites, using epi-fluorescence microscopy, but a general, broad-based application of this approach has not yet been described. Moreover, it should be emphasized that such an approach is limited to a detection of cellular injury, but does not describe the extent of the injury (lethal or not), or the ability of the cell to repair the damaged nucleic acids. It is also possible to distinguish between the DNA and RNA content of cells by using a specific staining combination such as Hoechst 33342 or DAPI (for DNA content) and Pyronin Y (for RNA content) [64], while general stains such as SYBR Green I and II or Acridine Orange have apparently different affinities or fluorescence emission for DNA and RNA. Additionally, RNA staining in pure cultures can be accomplished with FISH probes [33].

4.1.3 Applications

Detection of nucleic acids is a basic check to determine if bacterial cells have one of the essential components of life, but is typically used for viability assessment

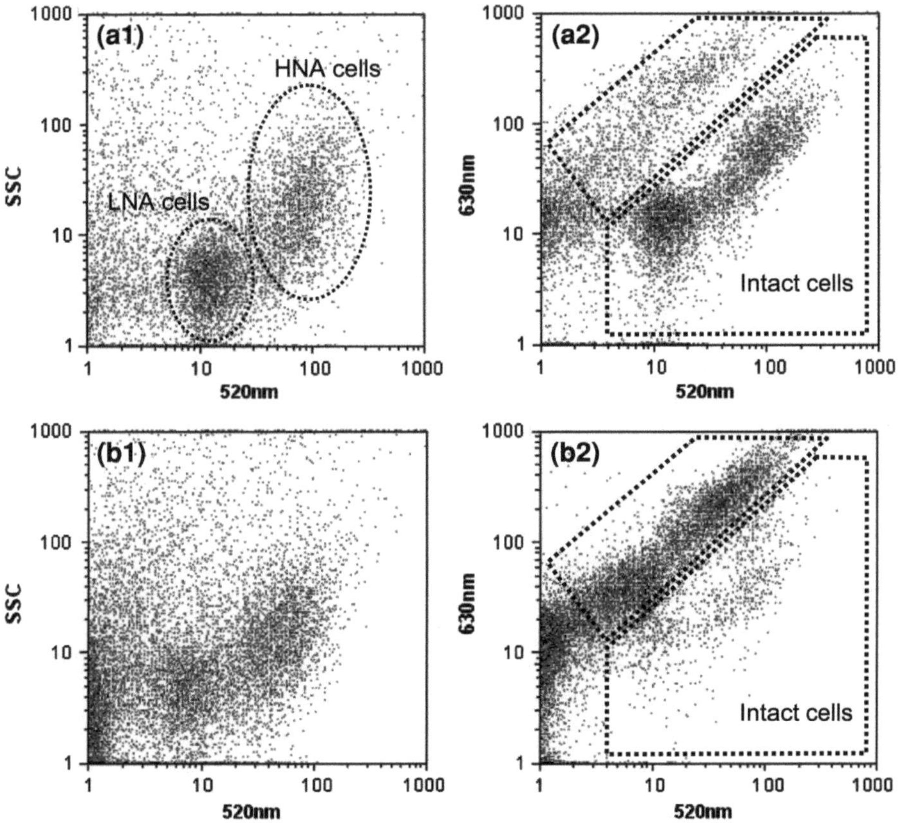

Fig. 5 Double-staining of indigenous high (HNA) and low (LNA) nucleic acid content bacteria from river water with SYBR Green I and propidium iodide before (**A1**, **A2**) and after (**B1**, **B2**) cell wall permeabilisation using a commercially available lysis buffer. Staining was done for 15 min at 25 °C; analysis was done with a Cyflow Space instrument (Partec)

only in situations where extremely aggressive damage to cells is expected. For example, assessment of non-specific damage to nucleic acids has been used to characterise disinfection of indigenous microbial communities during chlorination of drinking water [53, 54]. These authors have shown a considerable reduction in fluorescence intensity of cells stained with SYBR Green II after extensive chlorination, suggesting that this is linked to irreversible oxidative damage of cellular DNA and that this approach can be used as an indicator for complete disinfection. The efficacy of ozonation as disinfection process during drinking water treatment was similarly demonstrated with SYBR Green I and flow cytometry, which showed no detectable nucleic acids or cell structures after ozonation [23]. It has been proposed previously that the mere presence of specific clusters of high (HNA) and low (LNA) nucleic acid content bacteria, as visualised with common nucleic acid staining and flow cytometry (Fig. 5), differentiates between active and inactive bacteria in environmental samples [21, 39]. However, recent studies have directly questioned this interpretation (e.g., [69]), and it should be assumed that such a general approach is probably too simplistic, given the heterogeneity observed in indigenous microbial communities.

4.1.4 Opinion

Indigenous bacterial cells have widely different amounts of nucleic acids (both DNA and RNA), typified by the HNA and LNA bacteria found in nearly all natural water samples (Fig. 5) [69]. So the amount of cellular nucleic acids is per se not a useful indicator of life or death, especially when indigenous microbial communities are considered. While the complete lack of nucleic acids would be a clear indicator of cell death [63], it should be regarded as an extremely conservative approach to viability assessment. Firstly, only aggressive disinfection processes result in DNA damage that is detectable with general methods. Secondly, small but lethal damage to DNA (e.g., UV-C disinfection) would not be detected with this approach. The RNA content of bacteria is at best an indicator of cellular activity and not viability [60]. Moreover, when slow-growing or dormant cells are present in a sample, it is likely that RNA content of such cells is beneath the detection limit of most methods, while the cells remain essentially viable and even active.

4.2 Membrane Integrity

4.2.1 Principle

Bacterial membranes provide a highly regulated physical barrier between the intracellular and extracellular environment. Severe structural/physical damage to the cytoplasmic membrane of bacteria is usually irreversible and most likely leads to cell death [31]. An intact membrane can be detected through exclusion of molecules based on their molecular size, charge, hydrophobicity and presence of groups that cause steric hindrance for membrane diffusion [26, 63, 66].

4.2.2 Methodology

Membrane integrity staining was made famous by the unfortunately named "LIVE/DEAD" kit [10], containing propidium iodide (PI) as the selective molecule for permeabilised membranes. The name is "unfortunate" because it reflects hypothetical concepts of "life" or "death", rather than a methodological concept that is assessed (membrane permeability) [63], and results gained from the application of this method are often wrongly interpreted as meaning "alive or dead". PI is a large (668 Da), double-charged, red-fluorescent dye that intercalates with double-stranded DNA and normally only enters cells of which the cytoplasmic membrane of the cell is permeabilised [5, 10]. PI is commonly used in combination with a total nucleic acid stain such as SYBR Green I or SYTO 9 (Table 1; Fig. 5) [5, 19], and the specific flow cytometric patterns that can be detected in this manner are indicative of different degrees of cellular damage [5]. Several alternative stains are available, which provide the user with additional information of membrane damage. These include SYTOX Green, which is an asymmetrically triple-charged cyanine dye and ethidium homodimer-2 (Eth-D2) that consists of two phenanthridinium fluorophores and which is quadruply charged [36, 63]. In an interesting test, Schumann et al. [63] used these three stains together on natural aquatic communities and found decreasing dye permeability in the order of (1) PI, (2) SYTOX Green, and (3) Eth-D2, ascribing the result to the molecular structure of the dyes. These results suggest that the magnitude of membrane damage can be measured with the use of different dyes, but conclusive information in this regard is still required. Positive controls for membrane integrity staining include treatment with heat (90 °C, 3–5 min), ethanol, or with a membrane damaging detergent (Fig. 5).

4.2.3 Applications

Analysis of membrane damage is best used for viability assessment in situations where aggressive physico-chemical damage to cells is expected. This includes disinfection by heat [6], oxidants [40], several antibiotic compounds [51] and physical processes like sonication or electroperforation. PI has been used more than any other viability stain on both pure cultures and natural microbial communities from environmental samples [10, 19, 31]. For example, Berney et al. [6] showed with flow cytometric analysis that 70–80% of indigenous communities in drinking water displayed intact cytoplasmic membranes, while Schumann et al. [63] used epi-fluorescence microscopy to demonstrate that between 50 and 60% of indigenous surface water communities had intact membranes. However, these former examples are snapshot analysis of complex communities that reveals little about the processes leading to the membrane damage. From controlled laboratory scale experiments from the authors' group, Berney et al. [4] and Bosshard et al. [8] described die-off of *E. coli*, *Salmonella* and *Shigella* cells from solar (UV-A) radiation with a range of viability parameters, demonstrating that cell membrane damage is typically the last level of damage to become detectable. Similarly, Lisle et al. [40] described the impact of chlorination on *E. coli* cells with multiple parameters, and also showed that membrane integrity was the last viability indicator to be disabled in the disinfection process.

4.2.4 Opinion

Most membrane integrity stains are applicable for pure cultures as well as indigenous communities, and analysis with epi-fluorescence microscopy and flow cytometry is typically yielding clear results. Membrane integrity analysis is, similar to nucleic acid damage, a conservative indicator for viability [4, 40]. It can be assumed that a cell with a severely damaged cytoplasmic membrane can be considered as "dead", due to an inability to maintain a unique intracellular environment [5, 49], although some exceptions to this have been noted and discussed in some detail [47, 65]. However, the reverse argument (that cells with intact membranes are "alive") is not necessarily true [5, 31]. A straightforward example is UV-C disinfection (discussed above), which would result in inactivation of bacteria without any immediate detectable damage to the cytoplasmic membrane. It is of importance to note that the value of membrane integrity stains is not to measure "life" or "death", but to assess a particular location-specific damage to cells. Hence, both the application/purpose and the interpretation of data from membrane integrity analysis, as well as the manner and nomenclature in which such data are reported, should be considered with the utmost care.

4.3 Membrane Potential

4.3.1 Principle

As discussed in detail in Sect. 3 (above), only living cells are in theory capable of maintaining a membrane potential. An irreversible loss of membrane potential would expose the cell to potentially lethal stress factors (e.g., pH, salts) in its environment. The processes shown in Fig. 2 demonstrate that a loss of membrane potential would have profound impacts on several vital cellular functions and processes, eventually leading to cell death.

4.3.2 Methodology

The presence of a membrane potential selectively regulates the passage of lipophilic cationic and anionic charged molecules through the cytoplasmic membrane. Depending on the charge of the dye, such molecules can either accumulate in polarized (cationic dyes) or depolarised cells (anionic dyes) [11, 31]. For example, uptake of anionic bis-(1,3-dibutylbarbituric acid) tri-methine oxonol ($DiBAC_4(3)$), also known as BOX, is limited to cells that are depolarized, upon which non-specific binding to intracellular proteins occurs [4, 26]. Alternatively, Rhodamine 123 (Rh123) is a polar cationic dye with a single delocalised positive charge, which crosses polarized membranes and accumulates in viable cells [20, 51]. Additional membrane potential dyes include 3,3′-dihexyloxacarbocyanine

(DiOC_6(3)), 3,3′-diethyloxacarbocyanine (DiOC_2(3)) and 3,3′-dipropylthiadicarbocyanine (DiSC_3(5)) [11, 50, 51]. Anionic dyes like DiBAC_4(3) are typically used to obtain a yes/no answer about the membrane potential of the cell and are thus suitable as viability stains. Other stains like DiOC_2(3) have been used to accurately measure the membrane potential and changes thereof in bacterial cells on a single-cell level [50]. For many of the membrane potential dyes, a pre-treatment step with EDTA is often needed to permeabilize the outer membrane [6, 31], but this treatment may by itself affect the cell's membrane potential. One drawback is that these dyes are usually fluorescent as such, which means background fluorescence can be a problem when analysing such samples. A dilution or washing step can be useful to avoid this problem. The appropriate control for membrane potential staining is to use uncoupling agents such as carbonyl cyanide m-chlorophenyl hydrazone (CCCP) or ionophores (e.g., nigericin, valinomycin) [31, 47, 51], while heat-killed cells (3 min, 90 °C) serve as a usable absolute negative control [6].

4.3.3 Applications

While some antibiotic compounds are thought to specifically target membrane potential [51], this parameter is often used as a sensitive and general indicator of the viability state of bacteria. Multiple reports have investigated the use of membrane potential stains for the analysis of bacterial pure cultures during fed-batch fermentations, with the general conclusion that this method can provide fast and useful information for monitoring industrial processes [27, 70]. Novo et al. [51] have used a combination of membrane integrity and membrane potential dyes to analyse and describe the mode of action of a range of antibiotics on pure cultures. A typical example is streptomycin, which causes K^+ efflux and an inhibition of respiration in bacterial cells, both of which are factors that would contribute to membrane depolarisation. In contrast, chloramphenicol, which targets protein synthesis mechanisms, had no detectable short/mid-term impact on bacterial membrane potential [51]. The same authors also noted that fluorescence intensity of membrane potential stains (e.g. DiBAC_4(3)) can be influenced by cell size, which may hamper application and interpretation when natural communities are analysed. In this regard, Müller and Nebe-von Caron [46] note that a typical approach is to create a ratio between fluorescence intensity and cell size when membrane potential is analysed (see also [66]). The application of membrane potential dyes to natural communities has been largely limited [31]. In one such application, Berney et al. [6] have demonstrated the use of DiBAC_4(3) on indigenous drinking water communities, showing 60–80% of cells in non-chlorinated drinking water showing "viable" polarized membranes. This was, however, not in relation to specific bactericidal processes, but rather a general "snapshot" assessment of an indigenous community. For accurate application of membrane potential methods on natural communities, additional information is needed about the accuracy of staining small cells that are prevalent in natural environments, and the interaction of the stains with charged compounds in the water samples [31].

4.3.4 Opinion

Membrane potential is a rather sensitive viability parameter and during general disinfection processes (e.g., solar disinfection), the loss of the membrane potential occurs in conjunction with a failure in cellular energy systems (ATP formation and the energy-dependent uptake of molecules), often well before membrane permeabilisation occurs [4, 8]. As a result, extensively damaged cells that have permeable membranes (Sect. 4.2) would typically display a lack of membrane potential as well [6]. However, the loss of membrane potential as a result of cellular inactivation may be time-dependent. For example, many bacteriostatic agents work through interference with protein synthesis, and therefore an immediate response in membrane potential is not expected [51]. Hence, it is essential to assess not only the *presence* of membrane potential, but also the time-dependant *maintenance* thereof. Of the available methods, the anionic dyes seem to have broader applicability, but it should always be remembered that the experimental outcome is not a direct measurement of membrane potential, but rather the specific behaviour of a given dye in a cell. While extensive work has been done with membrane potential staining of pure cultures, the application in natural samples is largely limited [6, 20, 31].

4.4 Efflux Pump Activity

4.4.1 Principle

Some molecules that can cross intact cytoplasmic membranes by passive diffusion are pumped out of active cells via specific or non-specific proton antiport transport systems. The pumping is ATP-independent but directly dependent on a trans-membrane electrochemical gradient. A loss of this pumping activity is therefore indicative of a change in membrane potential or damage to the membrane.

4.4.2 Methodology

The most commonly used stain for efflux pump activity analysis is ethidium bromide (EB) [30, 31, 41]. The small EB molecules (394 Da) with a single positive charge enter the cell through passive uptake. If the membrane pumps are no longer functioning it accumulates in the cell and binds to nucleic acids. A common approach is the use of EB in combination with a green fluorescent nucleic acid stain (e.g. SYBR Green I), which allows dynamic detection of small changes and intermediate cell states [41]. An appropriate control for EB-staining has been described as using a combination of sodium azide (NaN_3) and Tween-20, which apparently halts EB extrusion without compromising the membrane integrity [48], while treatment with verapamil, and m-chlorophenyl hydrazone also inhibit efflux pump activity [46].

4.4.3 Application

Efflux pump activity has been used successfully in the monitoring of industrial fermentation processes with pure cultures of bacteria. For example, Looser et al. [41] demonstrated that the damage caused to *E. coli* membranes through the expression of a heterologous membrane protein could be followed most accurately with EB-staining, and this can be used for real-time monitoring of the process. In several studies the group of Hewitt [27, 70] have used a protocol combining EB with PI (membrane integrity) and $DiBAC_4(3)$ (membrane potential) to describe various sub-populations occurring during industrial fed-batch fermentations.

4.4.4 Opinion

The absence of a pumping activity suggests potential stressful conditions, but not necessarily cell death. Müller and Nebe-von Caron [46] suggested that active transport is obviously low in that are viable but with very low activity, which may lead to false interpretations of staining results. Nebe-von Caron et al. [48] as well as Looser et al. [41] described EB as more sensitive than either membrane potential dyes (e.g., $DiBAC_4(3)$) or membrane integrity dyes (e.g., PI) for the detection of physiological changes in cells, and a similar observation was also made for solar irradiated cultures [4]. However, while EB staining works fairly well for assessment of pure cultures, Joux and Lebaron [31] opined that the current methodology for detection of pumping activity is not universal enough for analysis of environmental samples—an opinion that is shared by the present authors. As discussed below, efflux pump activity can also have a negative impact on other staining methods (e.g., enzyme activity staining), where either the fluorochromes or the cleavage products are in some cases actively exported from viable cells.

4.5 Respiratory Activity

4.5.1 Principle

Respiratory activity in bacteria depends on a functioning electron transport chain, which is the main process for maintaining a membrane potential (Fig. 2). The presence of respiratory activity is therefore an indicator of viability in cells, linking membrane potential and the recycling of reducing equivalents (e.g. NAD+) that are produced in many catabolic reactions.

4.5.2 Methodology

Respiratory activity in bacteria can be detected by the use of artificial electron acceptors, specifically tetrazolium salts, which are reduced to insoluble formazan products [71]. The reduction of tetrazolium salts are indirectly linked with the enzymes that form part of the electron transport chain (NADH dehydrogenase I & II, succinate dehydrogenase, cytochrome c reductase and the terminal oxidases), which is why respiratory activity is also often described together with enzyme activity protocols [67]. The two most commonly used tetrazolium salts are INT (2-(*p*-iodophenyl)-3-(*p*-nitrophenyl)-5-phenyltetrazolium chloride), which is reduced to INT-formazan, and CTC (5-cyano-2,3-ditolyl tetrazolium chloride), which is reduced to red-fluorescent 3-cyan-1,5-ditolyl formazan (CTF) [71]. Of these two dyes, CTC seems to be the preferred choice based on the easy detection of the fluorescent product with fluorescence microscopy and/or flow cytometry. Different staining protocols include differences in staining times (4–24 h), the addition of intermediate electron carriers, changing the oxygen concentrations during staining, and also the inclusion of additional substrates in the medium during staining [14, 71]. For negative controls, the dissipation of the membrane potential can be achieved by the addition of a combination of nigericin and valinomycin, or similar reagents as described above in Sect. 4.3 [12, 57].

4.5.3 Application

Respiratory activity staining is one of the older viability assays, with already numerous applications in both pure cultures and environmental microbiology during the last two decades [59, 71]. For example, Schumann et al. [63] found reasonably good correlations between CTC reduction and esterase activity (discussed below) in a number of environmental samples. For a comprehensive review of various CTC applications and comparisons with other methods, the reader is referred to the comprehensive discussion in the study of Creach et al. [14].

4.5.4 Opinion

Considerable disagreement exists to the exact value of this method for the analysis of environmental samples [14, 68]. One of the shortcomings of the CTC assay is that it often requires long staining times (up to 24 h) [71], which makes it less interesting for a rapid microbiological assay. Such a long staining time could result in other viability changes (e.g., die-off, regrowth) in the samples. Also, for flow cytometric analysis it is essential that the crystals are formed inside the cells, which is not always the case. Moreover, it has been noted that not all bacteria are capable of reducing formazan salts (because not all bacteria have a functioning electron transport chain), which seriously questions the general applicability of the method on indigenous environmental communities [14, 68, 71].

4.6 Enzymatic Activity

4.6.1 Principle

Bacteria maintain a number of housekeeping enzymes with relative general functions (e.g., esterases, dehydrogenases, peptidases). The absence of these enzymes from a cell suggests an inability of the cell to synthesise and maintain new proteins, which is indicative of inactivity and potential death [49]. Enzyme activity is usually measured through the detection of specific cleavage products in cells.

4.6.2 Methodology

All bacteria contain relative unspecific esterases, and the presence thereof can be detected through the addition of an uncharged, non-fluorescent and lipophilic substrate, which upon cleavage, becomes a fluorescent polar product that is retained to varying degrees in cells. For this purpose, a very wide variety of substrates have been tested rather extensively, including fluorescein diacetate (FDA), carboxyfluorescein diacetate (CFDA), carboxyfluorescein diacetate acetoxymethyl ester (CFDA-AM), 2′,7′-bis-(2-carboxyethyl)-5-(and-6)-carboxyfluorescein-AM (BCECF-AM) and calcein-AM [11, 31, 36]. Some of the abovementioned dyes tend to be pH-dependent, which influences their efficacy [46]. Additionally, some of the fluorogenic substrates or cleavage products can be actively exported from bacterial cells [16, 46], again hampering the interpretation of staining data. As a result, various different staining protocols exist in which different substrates are favoured. It is, however, essential that any chosen protocol should be carefully controlled, e.g., by inhibition of efflux pumps (Sect. 4.4.2) and also by complete heat-inactivation of enzymes (negative control). Apart from esterases, at least two other enzyme systems are often targeted in viability assays. Firstly, the reduction of tetrazolium dyes (e.g. CTC, see section above) is indirectly linked to the enzymes of the electron transport chain (see Sect. 4.5.). Secondly, Schumann et al. [63] describe the use of 7-amino-4-chloromethylcoumarin, L-leucine amide hydrocloride (CMAC-Leu) for the measurement of peptidase activity in bacteria.

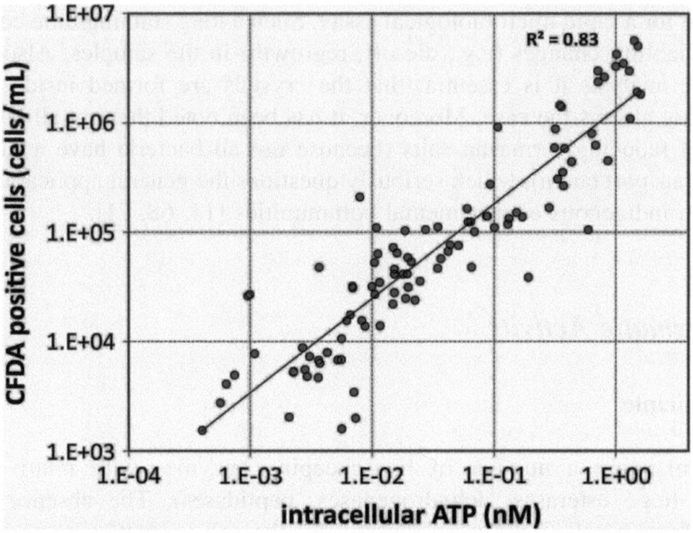

Fig. 6 A general correlation between the concentration of ATP and esterase positive cells (CFDA staining) in water samples ($n = 80$) including groundwater, tap water, surface water and wastewater effluent. Samples were analysed as described in Berney et al. [6]

4.6.3 Application

Specific disinfection processes like heat-killing would directly affect enzymatic activity in bacterial cells through denaturation of the enzymes. Therefore, loss of specific enzyme activity can be indicative of a specific location or type of cell damage (e.g., proteins). Schumann et al. [63] assessed both esterase and peptidase activity in a number of natural surface water samples. In general, higher esterase activity was detected in all samples, but the exact interpretation from studies based on such grab-samples is difficult to make. In a recent study in the authors' research group, esterase activity was measured along with several other viability parameters in different drinking water samples. What was particularly interesting was that a relatively good correlation was found between the concentration of esterase positive cells and the concentration of ATP in the water samples (Fig. 6). However, as discussed below, there is sufficient reason to believe that esterase data should be regarded with care, particularly during disinfection experiments.

4.6.4 Opinion

Most of the substrates used for enzyme activity measurements enter bacterial cells through passive diffusion. Moreover, the actual cleavage (substrate–enzyme reaction) is independent of cellular energy [49]. Hence, the detection of enzyme activity does not necessarily suggest cell viability, and recently killed cells are more than likely to still display enzymatic activity. In fact, Breeuwer and Abee [11] reported several cases where dead cells displayed enhanced esterase activity, probably due to better transport of the dyes into dead cells while enzymes remain active. Therefore, a key aspect again is the time-dependent *maintenance* of enzymatic activity in cells, rather than the mere presence thereof in a snapshot analysis of a sample. On the other hand, the absence of enzymatic activity is by no means an indicator of dead cells, but can also be associated with a low activity state in bacteria. Moreover, Diaz et al. [16] describes problems with dye uptake as well as active dye extrusion. Heat-killing is an often-used control for esterase stains [6, 55]. However, it has to be recognized that heat would denature proteins, thereby damaging and inactivating the esterase enzymes. Heat-killing is therefore only a control for enzyme presence and activity, but it is not a control for the response of enzyme activity relative to cell viability (or the lack thereof). Hence, it is the authors' opinion that enzymatic activity should be considered with care, and reported accurately to the experimental or environmental conditions that were studied.

4.7 Cellular Energy: Adenosine Tri-Phosphate (ATP)

4.7.1 Principle

All microorganisms require a minimum amount of energy to maintain life-sustaining processes. Energy in bacteria is generally present in the form of ATP, which can be generated either through oxidative phosphorylation (by creating a proton motive force that drives the ATP synthase) or substrate level phosphorylation (e.g., glycolysis). Additionally, bacteria can accumulate compounds like phosphoenolpyruvate, glycogen or triacylglyceride, which can be converted to ATP under energy starvation conditions. ATP is commonly referred to as the energy currency of microbial cells [29]. It is turned over rapidly in cells due to the coupling of anabolic and catabolic reactions, and as a result, ATP concentrations often respond to physiological states in microorganisms.

4.7.2 Methodology

Khlyntseva et al. [35] presents an extensive review of ATP determination methods (not limited only to bacterial cells). Although bacterial ATP measurements on single-cell level are not commonly used [29], bulk ATP analysis has evolved during the last decade into easy and straightforward measurements [24]. In standard laboratory applications, ATP measurements are typically accomplished with commercially available ATP kits, based on detection of luminescence generated from the interaction of extracted ATP with the luciferin–luciferase enzyme complex. Intracellular and extracellular ATP can be distinguished from each other by using 0.1 μm filtration.

4.7.3 Application

In a recent paper from the authors' group, it was demonstrated that a good general correlation was obtained between the concentration of intact cells (measured with SGPI-staining) and the concentration of bacterial ATP for a variety of indigenous aquatic microbial communities [24]. Similarly, a good correlation was obtained with esterase-positive cells following CFDA-staining (Fig. 6). However, it should be emphasized that this data set looked specifically at freshwater environments where no dramatic stress conditions could be expected. During sunlight disinfection experiments, it was demonstrated that ATP is sensitive to cellular stress and often displayed decreases considerably earlier than parameters such as cultivability and substrate uptake [4, 8].

4.7.4 Opinion

The advantage of ATP analysis in the context of viability is that it provides the user with an independent tool to compliment the flow cytometry or microscopy data (Figs. 3, 6). Unfortunately, this is in most applications only a bulk parameter and not a single-cell measurement. This is indeed problematic, since the ATP content of bacterial cells is not constant. For example, HNA cells contain on average about 10-fold more ATP than LNA cells [69]. This means that bulk ATP measurements can easily lead to erroneous conclusions when the data are compared to single-cell data. Also, ATP measurements do not take into account the rapid changes in cellular ATP levels, and specific responses to environmental conditions, e.g., increased catabolism or decreased anabolism. It is a known fact that the energy in a cell depends on the balance between the various adenosine phosphates [ATP, adenosine di-phosphate (ADP) and adenosine monophosphate (AMP)] [35]. In addition, it was shown that extracellular ATP can have an influence on total ATP measurements, and should be considered during analysis [24].

5 Conclusions

It is highly unlikely that any single silver-bullet viability staining method exist, due to the heterogeneous nature of microbial life. As a result, the best approach is the use of a combination of cultivation-independent methods that target different cellular processes linked to viability in order to gain specific and meaningful information on different cellular states, types of cell damage and the degree of cellular injury. It is emphasized that the methods should be carefully chosen and optimised prior to application, bearing in mind the particular environment and organisms that are analysed. In this regard, critical points to consider are (1) the cellular process measured with the method (Fig. 3), (2) the specificity of the method towards the target organism, (3) the inherent link between different methods/processes (e.g., Figs. 3, 6), (4) the sequence and time-dependent manner in which cellular processes respond to a specific stress (e.g., Fig. 1), and the ability of cells to recover from a certain injury. The application of cultivation-independent viability assays to environmental samples is a needed and worthwhile endeavour in order to understand bacterial behaviour in complex natural environments. However, pure culture work, specifically using defined organisms under diverse (but controlled) stress conditions, coupled to cell sorting and re-cultivation, has the potential to enhance the actual understanding and interpretation of cultivation-independent methods [46, 49].

References

1. Al-Adhami BH, Nichols RAB, Kusel JR, O'Grady JO, Smith HV (2007) Detection of UV-induced thymine dimers in individual *Cryptosporidium parvum* and *Cryptosporidium hominis* oocysts by immunofluorescence microscopy. Appl Environ Microbiol 73:947–995
2. Asano S, Iijima K, Suzuki K, Motoyama Y, Ogata T, Kitagawa Y (2009) Rapid detection and identification of beer-spoilage lactic acid bacteria by microcolony method. J Biosci Bioeng 108:124–129
3. Ateya DA, Erickson JS, Howell PB, Hilliard LR, Golden JP, Ligler FS (2008) The good, the bad, and the tiny: a review of microflow cytometry. Anal Bioanal Chem 391:1485–1498
4. Berney M, Weilenmann HU, Egli T (2006) Flow-cytometric study of vital cellular functions in *Escherichia coli* during solar disinfection (SODIS). Microbiol-SGM 152:1719–1729
5. Berney M, Hammes F, Bosshard F, Weilenmann HU, Egli, T (2007) Assessment and interpretation of bacterial viability by using the LIVE/DEAD BacLight Kit in combination with flow cytometry. Appl Environ Microbiol 73:3283–3290
6. Berney M, Vital M, Hülshoff I, Weilenmann HU, Egli T, Hammes F (2008) Rapid, cultivation-independent assessment of microbial viability in drinking water. Water Res 42:4010–4018
7. Bogosian G, Bourneuf EF (2001) A matter of bacterial life and death. EMBO Rep 2:770–774
8. Bosshard F, Berney M, Scheifele M, Weilenmann HU, Egli T (2009) Solar disinfection (SODIS) and subsequent dark storage of *Salmonella typhimurium* and *Shigella flexneri* monitored by flow cytometry. Microbiol-SGM 155:1310–1317
9. Bosshard F, Riedel K, Schneider T, Geiser C, Bucheli M, Egli T (2010) Protein oxidation and aggregation in UVA-irradiated *Escherichia coli* cells as signs of accelerated cellular senescence. Environ Microbiol. doi:10.1111/j.1462-2910.2010.02268.x
10. Boulos L, Prevost M, Barbeau B, Coallier J, Desjardins R (1999) LIVE/DEAD(R) BacLight (TM): application of a new rapid staining method for direct enumeration of viable and total bacteria in drinking water. J Microbiol Methods 37:77–86
11. Breeuwer P, Abee T (2000) Assessment of viability of microorganisms employing fluorescence techniques. Int J Food Microbiol 55:193–200
12. Breeuwer JA, Abee T (2004) Assessment of the membrane potential, intracellular pH and respiration of bacteria employing fluorescence techniques. In: Kowalchuk GA, de Bruijn FJ, Head IM, Akkermans AD, van Elsas JD (eds) Molecular microbial ecology manual, 2nd edn. Springer, pp 1563–1580. ISBN: 978-1-4020-2173-3

13. Brehm-Stecher B, Johnson EA (2004) Single-cell microbiology: tools, technologies, and applications. Microbiol Mol Biol Rev 68:538–559
14. Creach V, Baudoux AC, Bertru G, Le Rouzic B (2003) Direct estimate of active bacteria: CTC use and limitations. J Microbiol Methods 52:19–28
15. Czechowska K, Johnson DR, van der Meer JR (2008) Use of flow cytometric methods for single-cell analysis in environmental microbiology. Curr Opin Microbiol 11:205–212
16. Diaz M, Herrero M, Garcia LA, Quiros C (2010) Application of flow cytometry to industrial microbial bioprocesses. Biochem Eng J 48:385–407
17. Dimroth P, Cook GM (2004) Bacterial Na^+- or H^+-coupled ATP synthases operating at low electrochemical potential. Adv Microb Physiol 49:175–218
18. Dubelaar GBJ, Gerritzen PL (2000) CytoBuoy: a step forward towards using flow cytometry in operational oceanography. Sci Mar 64:255–265
19. Falcioni T, Papa S, Gasol JM (2008) Evaluating the flow-cytometric nucleic acid double-staining protocol in realistic situations of planktonic bacterial death. Appl Environ Microbiol 74:1767–1779
20. Gasol JM, del Giorgio PM (2000) Using flow cytometry for counting natural planktonic bacteria and understanding the structure of planktonic bacterial communities. Sci Mar 64:197–224
21. Gasol JM, Zweifel UL, Peters F, Fuhrman JA, Hagstrom A (1999) Significance of size and nucleic acid content heterogeneity as measured by flow cytometry in natural planktonic bacteria. Appl Environ Microbiol 65:4475–4483
22. Hammes F, Egli T (2010) Cytometric methods for measuring bacteria in water: advantages, pitfalls and applications. Anal Bioanal Chem 397:1083–1095
23. Hammes F, Berney M, Wang Y, Vital M, Köster O, Egli T (2008) Flow-cytometric total bacterial cell counts as a descriptive microbiological parameter for drinking water treatment processes. Water Res 42:269–277
24. Hammes F, Goldschmidt F, Vital M, Wang Y, Egli T (2010) Measurement and interpretation of microbial adenosine tri-phosphate (ATP) in aquatic environments. Water Res 44:3915–3923
25. Harris GD, Adams VD, Sorensen DL, Curtis MS (1987) Ultraviolet inactivation of selected bacteria and viruses with photoreactivation of the bacteria. Water Res 21:687–692
26. Hewitt CJ, Nebe-von Caron G (2004) The application of multi-parameter flow cytometry to monitor individual microbial cell physiological state. Adv Biochem Eng Biotechnol 89:197–223
27. Hewitt CJ, Nebe-Von Caron G, Axelsson B, McFarlane CM, Nienow AW (2000) Studies related to the scale-up of high-cell-density *E. coli* fed-batch fermentations using multiparameter flow cytometry: effect of a changing microenvironment with respect to glucose and dissolved oxygen concentration. Biotechnol Bioeng 70:381–390
28. Hoefel D, Grooby WL, Monis PT, Andrews S, Saint CP (2003) Enumeration of water-borne bacteria using viability assays and flow cytometry: a comparison to culture-based techniques. J Microbiol Methods 55: 585–597
29. Imamura H, Nhat KP, Togawa H, Saito K, Iino R, Kato-Yamada Y, Nagai T, Noji HKP (2009) Visualization of ATP levels inside single living cells with fluorescence resonance energy transfer-based genetically encoded indicators. Proc Natl Acad Sci USA 106:15651–15656
30. Jarnagina JL, Luchsingera DW (1980) The use of fluorescein diacetate and ethidium bromide as a stain for evaluating viability of *Mycobacteria*. Biotechnic and Histochemistry 55:253–258
31. Joux F, Lebaron P (2000) Use of fluorescent probes to assess physiological functions of bacteria at single-cell level. Microbes Infect 2:1523–1535
32. Jung H (2001) Towards the molecular mechanism of Na(+)/solute symport in prokaryotes. Biochim Biophys Acta 1505:131–143

33. Karner M, Fuhrman JA (1997) Determination of active marine bacterioplankton: a comparison of universal 16S rRNA probes, autoradiography, and nucleoid staining. Appl Environ Microbiol 63:1208–1213
34. Kell DB, Kaprelyants AS, Weichart DH, Harwood CR, Barer MR (1998) Viability and activity in readily culturable bacteria: a review and discussion of the practical issues. Antonie Leeuwenhoek 73:169–187
35. Khlyntseva SV, Bazel YR, Vishnikin AB, Andruch V (2009) Methods for the determination of adenosine triphosphate and other adenine nucleotides. J Anal Chem 64:657–673
36. King MA (2000) Detection of dead cells and measurement of cell killing by flow cytometry. J Immunol Methods 243:155–166
37. Laussermair E, Schwarz E, Oesterhelt D, Reinke H, Beyreuther K, Dimroth P (1989). The sodium ion translocating oxaloacetate decarboxylase of *Klebsiella pneumoniae*. Sequence of the integral membrane-bound subunits beta and gamma. J Biol Chem 264:14710–14715
38. Lebaron P, Parthuisot N, Catala P (1998) Comparison of blue nucleic acid dyes for flow cytometric enumeration of bacteria in aquatic systems. Appl Environ Microbiol 64:1725–1730
39. Lebaron P, Servais P, Agogue H, Courties C, Joux F (2001) Does the high nucleic acid content of individual bacterial cells allow us to discriminate between active cells and inactive cells in aquatic systems? Appl Environ Microbiol 67:1775–1782
40. Lisle JT, Pyle BH, McFeters GH (1999) The use of multiple indices of physiological activity to access viability in chlorine disinfected *Escherichia coli* O157:H7. Lett Appl Microbiol 29:42–47
41. Looser V, Hammes F, Keller M, Berney M, Kovar K, Egli T (2005) Flow-cytometric detection of changes in the physiological state of *E. coli* expressing a heterologous membrane protein during carbon-limited fedbatch cultivation. Biotechnol Bioeng 92:69–78
42. Lunde CS, Hartouni SR, Janc JW, Mammen M, Humphrey PP, Benton BM (2009) Telavancin disrupts the functional integrity of the bacterial membrane through targeted interaction with the cell wall precursor lipid II. Antimicrob Agents Chemother 53:3375–3383
43. Madigan MT, Martinko JM (2006) Brock biology of microorganisms, 11th edn. Pearson Benjamin Cummings, San Francisco
44. Malacrino P, Zapparoli G, Torriani S, Dellaglio F (2001) Rapid detection of viable yeasts and bacteria in wine by flow cytometry. J Microbiol Methods 45:127–134
45. McFeters GA, Yu FP, Pyle BH, Stewart PS (1995) Physiological assessment of bacteria using fluorochromes. J Microbiol Methods 21:1–13
46. Müller S, Nebe-von Caron G (2010) Functional single-cell analyses–flow cytometry and cell sorting of microbial populations and communities. FEMS Microbiol Rev. doi: 10.1111/j.1574-6976.2010.00214.x
47. Müller S, Harms H, Bley T (2010) Origin and analysis of microbial population heterogeneity in bioprocesses. Curr Opin Biotechnol 21:100–113
48. Nebe-von Caron G, Stephens P, Badley RA (1998) Assessment of bacterial viability status by flow cytometry and single cell sorting. J Appl Microbiol 84:988–998
49. Nebe-von Caron G, Stephens PJ, Hewitt CJ, Powell JR, Badley RA (2000) Analysis of bacterial function by multi-colour fluorescence flow cytometry and single cell sorting. J Microbiol Methods 42:97–114
50. Novo DJ, Perlmutter NG, Hunt RH, Shapiro HM (1999). Accurate flow cytometric membrane potential measurement in bacteria using diethyloxacarbocyanine and a ratiometric technique. Cytometry 35:55–63
51. Novo DJ, Perlmutter NG, Hunt RH, Shapiro HM (2000) Multiparameter flow cytometric analysis of antibiotic effects on membrane potential, membrane permeability, and bacterial counts of *Staphylococcus aureus* and *Micrococcus luteus*. Antimicrob Agents Chemother 44:827–834
52. Oliver JD (2005) The viable but nonculturable state in bacteria. J Microbiol 43:93–100
53. Phe MH, Dossot M, Guilloteau H, Block JC (2005) Nucleic acid fluorochromes and flow cytometry prove useful in assessing the effect of chlorination on drinking water bacteria. Water Res 39:3618–3628

54. Phe MH, Dossot M, Guilloteau H, Block JC (2007) Highly chlorinated *Escherichia coli* cannot be stained with propidium iodide. Can J Microbiol 53:664–670
55. Porter J, Deere D, Pickup R, Edwards C (1996) Fluorescent probes and flow cytometry: new insights into environmental bacteriology. Cytometry 23:91–96
56. Postgate JR (1969) Viable counts and viability. In: Norris JR, Ribbons DW (eds) Methods in microbiology, vol 1. Academic Press Inc., London
57. Rao M, Streur TL, Aldwell FE, Cook GM (2001) Intracellular pH regulation by *Mycobacterium smegmatis* and *Mycobacterium bovis* BCG. Microbiology 147:1017–1024
58. Rappé M, Giovannoni S (2003) The uncultured microbial majority. Annu Rev Microbiol 57:369–394
59. Rodriquez GG, Phipps D, Ishiguro K, Ridgway HF (1992) Use of a fluorescent redox probe for direct visualisation of actively respiring bacteria. Appl Environ Microbiol 58:1801–1808
60. Roszak DB, Colwell RR (1987) Survival strategies of bacteria in the natural environment. Microbiol Rev 51:365–379
61. Saint-Ruf C, Cordier C, Megret J, Matic I (2010) Reliable detection of dead microbial cells using fluorescent hydrazides. Appl Environ Microbiol 76:1674–1678
62. Schloss PD, Handelsman J (2004) Status of the microbial sensus. Microbiol Mol Biol Rev 68:686–691
63. Schumann R, Schiewer U, Karsten U, Rieling T (2003) Viability of bacteria from different aquatic habitats. II. Cellular fluorescent markers for membrane integrity and metabolic activity. Aquat Microb Ecol 32:137–150
64. Shapiro HM (1981) Flow cytometric estimation of DNA and RNA content in intact cells stained with Hoechst 33342 and Pyronin Y. Cytometry 2:143–150
65. Shi L, Gunther S, Hubschmann T, Wick LY, Harms H, Müller S (2007) Limits of propidium iodide as a cell viability indicator for environmental bacteria. Cytometry A 71:592–598
66. Sträuber H, Müller S (2010) Viability states of bacteria–specific mechanisms of selected probes. Cytometry A 77:623–634
67. Tracy BP, Gaida SM, Papoutsakis ET (2010) Flow cytometry for bacteria: enabling metabolic engineering, synthetic biology and the elucidation of complex phenotypes. Curr Opin Microbiol 21:85–99
68. Ullrich S, Karrasch B, Hoppe HG (1999) Is the CTC dye technique an adequate approach for estimating active bacterial cells? Aquat Microb Ecol 17:207–209
69. Wang Y, Hammes F, Boon N, Chami M, Egli T (2009) Isolation and characterization of low nucleic acid (LNA)-content bacteria. ISME J 3:889–902
70. Want A, Thomas ORT, Kara B, Liddel J, Hewitt CJ (2009) Studies related to antibody fragment (Fab) production in *Escherichia coli* W3110 fed-batch fermentation processes using multi-parameter flow cytometry. Cytometry A 75:148–154
71. Winding A, Binnerup SJ, Sorensen J (1994) Viability of indigenous soil bacteria assayed by respiratory activity and growth. Appl Environ Microbiol 60:2869–2875
72. Ziglio G, Andreottola G, Barbesti S, Boschetti G, Bruni L, Foladori P, Villa R (2002) Assessment of activated sludge viability with flow cytometry. Water Res 36:460–468

Adv Biochem Engin/Biotechnol (2011) 124: 151–181
DOI: 10.1007/10_2010_82

Published Online: 11 November 2010

Resolution of Natural Microbial Community Dynamics by Community Fingerprinting, Flow Cytometry, and Trend Interpretation Analysis

Petra Bombach, Thomas Hübschmann, Ingo Fetzer, Sabine Kleinsteuber, Roland Geyer, Hauke Harms and Susann Müller

Abstract Natural microbial communities generally have an unknown structure and composition because of their still not yet cultivable members. Therefore, understanding the relationships among the bacterial members, prediction of their behaviour, and controlling their functions are difficult and often only partly successful endeavours to date. This study aims to test a new idea that allows to follow community dynamics on the basis of a simple concept. Terminal restriction fragment length polymorphism (T-RFLP) analysis of bacterial 16S ribosomal RNA genes was used to describe a community profile that we define as composition of a community. Flow cytometry and analysis of DNA contents and forward scatter characteristics of the single cells were used to describe a community profile, which we define as structure of a community. Both approaches were brought together by a non-metric multidimensional scaling (n-MDS) for trend interpretation of changes in the complex community data sets. This was done on the basis of a graphical evaluation of the cytometric data, leading to the newly developed Dalmatian plot tool, which gave an unexpected insight into the dynamics of the unknown bacterial members of the investigated natural microbial community. The approach presented here was compared with other techniques described in the literature. The microbial community investigated in this study was

---

P. Bombach, T. Hübschmann, I. Fetzer, and S. Kleinsteuber are contributing authors to the same extent.

P. Bombach
Department of Isotope Biogeochemistry, UFZ, Helmholtz Centre for Environmental Research, Permoserstr. 15, 04318 Leipzig, Germany

T. Hübschmann, I. Fetzer, S. Kleinsteuber, H. Harms and S. Müller (✉)
Department of Environmental Microbiology, UFZ, Helmholtz Centre for Environmental Research, Permoserstr. 15, 04318 Leipzig, Germany
e-mail: susann.mueller@ufz.de

R. Geyer
Tecan Trading AG, Seestrasse 103, 8708 Männedorf, Switzerland

obtained from a BTEX contaminated anoxic aquifer. The indigenous bacteria were allowed to colonise in situ microcosms consisting of activated carbon. These microcosms were amended with benzene and one of the electron acceptors nitrate, sulphate or ferric iron to stimulate microbial growth. The data obtained in this study indicated that the composition (via T-RFLP) and structure (via flow cytometry) of the natural bacterial community were influenced by the hydro-geochemical conditions in the test site, but also by the supplied electron acceptors, which led to distinct shifts in relative abundances of specific community members. It was concluded that engineered environments can be successfully monitored by single cell analytics in combination with established molecular tools and sophisticated statistical analyses, a combination that holds great promise for studying and monitoring natural microbial community behaviour.

Keywords Bioprocess control · Biostimulation · Microbial community analysis · Microbial flow cytometry · In situ microcosms

Contents

1 Resolving Complex Microbial Communities

It has been a long-time ambition of scientists to link certain environmental processes to activities of individual cells within complex microbial communities. Studying solely variations of chemical parameters in the environment provides only limited information on an ecosystem. To unlock this "black box", traditional concepts focussed on isolation and subsequent characterisation of cultured microorganisms. Since most bacteria have not been cultivable until now, this concept does not seem appropriate for revealing the role of individual microorganisms within complex microbial communities.

Recently established methods, such as microautoradiography-fluorescence in situ hybridisation (MAR-FISH), stable isotope probing (SIP) of nucleic acids or quantitative PCR detecting key genes of biochemical pathways or their transcripts (for overview see Sect. 5.3 of this review and [45]), can provide information on the microbial communities and their catabolic functions, but these methodologies are often too elaborate and difficult to apply to field samples. During the last few years, population-correlated approaches on the basis of fluorescent labeling of single cells combined with cell sorting and further analysis of the sorted sub-populations have emerged as a valuable and robust tool to understand the development of intra-population functional variability [52]. The same basic idea can be used to resolve quantitatively natural microbial communities by labelling either already known bacteria (using phylogenetic probes for in situ hybridisation; [29, 30, 38]) or all bacteria with fluorescent dyes to facilitate cell sorting and subsequent analyses of the cells' functional potential by phylogenetic analysis (for overview see [33]). The easiest way to label all bacteria in a community is to stain the nucleic acids, which is widely done by the application of Syto® dyes resulting in the detection of so-called low nucleic and high nucleic acid (LNA and HNA, respectively) bacteria. HNA bacteria can contain high amounts of DNA and RNA, and are often regarded as very active, whereas LNA bacteria have been suggested to be generally inactive. However, recently very small representatives (below 0.8 μm, [51]) were found to be part of the LNA sub-community and proven to multiply. Microbial community dynamics can be followed at higher resolution when labeling exclusively the DNA. The blue fluorescent dye DAPI can be used for this purpose. Resulting DNA-pattern distributions and sorting of subsets of cells represent a quick and already well-established tool to allow investigators to observe natural microbial communities in detail [12, 19, 32]. The same approach is the basis for the routine in situ monitoring of complex groundwater microbial communities of unknown composition and structure, the outcome of which is presented in this study.

2 Investigation of In Situ Biodegradation as a Case Study

As a result of human activities, toxic chemicals have become ubiquitous contaminants of soils and groundwater worldwide. In 2006, the European

Environment Agency reported that approximately 250,000 contaminated field sites are in need of remediation due to the risks that pollutants pose to human and ecosystem health [8]. A similar situation exists in the USA, where there are approximately 294,000 contaminated field sites [46]. One frequent group of contaminants worldwide is made up of the aromatic hydrocarbons benzene, toluene, ethylbenzene and xylenes, collectively known as BTEX. BTEX compounds are petroleum products that penetrate into the groundwater as a result of leaking pipes and underground fuel tanks. Groundwater contamination by benzene is of particular concern, since benzene is highly water soluble and has been determined to be toxic and carcinogenic [2], thus affecting the quality of drinking water resources and function of ecosystems.

Microbial degradation of BTEX compounds is an important process for natural attenuation (NA) at field sites, leading to the removal of contaminants by mineralising them to carbon dioxide or methane. Thus, investigations of in situ biodegradation have become a highly relevant field of research within the field of aquifer decontamination. Previous studies have shown that BTEX compounds are amenable to microbial degradation under oxic [47] as well as under anoxic conditions using nitrate, sulphate, iron or carbon dioxide as electron acceptors (for review see [9, 43]). However, field studies revealed that anaerobic biodegradation was occurring at some sites, but failed at others [25]. This could be due to the geochemical heterogeneity of the contaminated sites, the associated varying potential of the indigenous microorganisms to degrade contaminants under anoxic conditions within a certain time frame, and the prevailing terminal electron acceptor processes influencing the energy yield [53]. Stimulation of the indigenous microbial communities is possible by addition of suitable electron acceptors and constitutes a conceivable bioremediation strategy called enhanced natural attenuation (ENA). To investigate the influence of specific electron acceptors, laboratory microcosms and enrichment cultures have often been used [26, 44]. The relevance of these studies for the actual in situ processes might be questioned as such laboratory settings cannot mirror environmental variables and reproduce the complexity of environmental habitats [28]. In situ microcosms, however, provide a valuable tool to study stimulation of microbial communities directly in the field. The microcosms [e.g., stainless steel cages filled with Bio-Sep® beads (BAC-TRAP®)] can be a useful prerequisite for a successful assessment of bioremediation strategies. During the deployment of these microcosms in a groundwater well, indigenous microorganisms colonise the Bio-Sep® beads, thus supporting development of microbial communities. This type of in situ microcosm has been used in several studies [37] to provide evidence for the in situ biodegradation of BTEX compounds, mono-chlorobenzene and MTBE (for a review see [5]), as well as to enrich a toluene-degrading microbial consortium for further cultivation experiments [4].

In this study, a combined approach of community fingerprinting, flow cytometry, and trend interpretation analysis was applied to follow and interprete microbial groundwater community microcosm colonisation. Therefore, in situ microcosms were amended with benzene and one of the electron acceptors nitrate, ferric iron or sulphate to mimic ENA treatment. The microcosms were exposed to three monitoring wells of a BTEX-contaminated aquifer to allow colonisation of the matrix by indigenous bacteria. The monitoring wells differed in their hydro-geochemical conditions with regard to benzene concentrations and electron acceptor availabilities. The colonised microbial communities were characterised according to features we defined as composition and structure. Community composition comprises phylogenetic information obtained by T-RFLP profiles, and community structure gives cytometric information on the single cells' DNA contents and light-scatter characteristics. Prominent sub-communities were separated by cell sorting for detailed phylogenetic analysis. As an outcome huge data sets were obtained, which could not be evaluated and compared by simple statistical approaches. Therefore, cytometric and phylogentic data were inspected in a combined approach by using the Dalmatian plot tool, which was newly developed for this study on the basis of non-metric multidimensional scaling (n-MDS) for trend interpretation of changes in the complex community data sets.

3 Experimental Procedures

3.1 Field Site

The BTEX-contaminated aquifer is located in the area of a former coal hydrogenation and benzene production plant close to Zeitz (Saxony-Anhalt, Germany). The main contaminant of the aquifer is benzene, with total concentrations of up to 1,000 mg l^{-1}. A detailed description of the history and hydro-geochemical conditions of the field site was given in previous articles [41, 49]. To assess the geology and contaminant distribution at the field site, a very dense monitoring network has been installed with over 90 monitoring wells. The microcosm experiments were performed in the wells 18/00, 7/99, and 52/03, located in the upper aquifer along a transect with decreasing contaminant concentration. Groundwater samples for hydro-geochemical analysis were taken 2 days before microcosm deployment. An overview on the hydro-geochemical parameters is given in Table 1.

3.2 Exposure of In Situ Microcosms

The field experiment was performed using BACTRAP®s, a microcosm system consisting of perforated stainless steel cages of 10 cm × 5 cm × 0.5 cm filled with 1 g Bio-Sep® beads (kindly provided by Kerry Sublette, University of Tulsa, OK). The spherical beads with a diameter of 2–3 mm were generated from a composite of 75% powdered activated carbon and 25% aramid polymer (Nomex) [37]. The beads have a porosity of 75%, an internal surface area of >600 m^2 g^{-1}

Table 1 Concentrations of benzene, nitrate (NO_3^-), sulphate (SO_4^{2-}), manganese (Mn), iron (Fe), nitrite (NO_2^-), ammonium (NH_4^+) and sulphide (S^{2-}) in the groundwater wells Zz 7/99, Zz 18/00 and Zz 52/03

Well	Benzene (mg l^{-1})	Redox potential (mV)	NO_3^- (mg l^{-1})	SO_4^{2-} (mg l^{-1})	Mn (mg l^{-1})	Fe (mg l^{-1})	NO^{2-} (mg l^{-1})	NH_4^+ (mg l^{-1})	S^{2-} (mg l^{-1})
Zz 7/99	179	−203	n.d.	2.40	0.57	2.72	0.02	2.29	0.89
Zz 18/00	241	−267	n.d.	9.10	3.26	42.40	0.05	5.43	5.23
Zz 52/03	3.72	+62	41.20	462	2.18	0.23	0.06	17.74	0.04

n.d. Nitrate concentration under the detection limit of 0.22 mg l^{-1}

and surface pore sizes of 1–10 μm, thus enabling microorganisms to penetrate. A detailed description of the BACTRAP system is given elsewhere [10, 18, 37].

In this study, five different types of BACTRAP® microcosms were prepared for each monitoring well: microcosms amended with benzene, benzene and nitrate, benzene and sulphate, benzene and ferric iron (in the following this is referred to iron for simplification), respectively, and microcosms without substrate and electron acceptor as controls. For preparation of microcosms, Bio-Sep® beads were heated to 300°C for 4 h to remove organic residues and were filled in perforated stainless steel cages. After sterilisation and hydration, the microcosms were each loaded with 133 μl benzene via the gas phase under reduced pressure as described previously [10]. The amendment with the electron acceptors was performed with $NaNO_3$, Na_2SO_4, and $FeCl_3$ to a concentration of 412 ± 8 mg g^{-1}, 472 ± 24 mg g^{-1}, and 490 ± 2 mg g^{-1}, respectively, during the manufacturing of the beads. All microcosms were stored in anoxic sterile water during their transport to the field site. The five different types of microcosms were fixed to a stainless packer system with 5 cm distance to each other and deployed 5 m below the groundwater table in the wells 18/00, 7/99, and 52/03.

3.3 Cell Preparation and DNA Staining

After 42 days of exposure, all microcosms were recovered, and the beads of each microcosm were transferred to a glass tube. The beads were immediately crushed mechanically and treated with 5 ml fixation buffer (0.5 mM sodium chloride, 15 mM sodium molybdate). Hydrogen peroxide (125 mM final concentration) and sodium azide (15 mM final concentration) were added immediately, and the samples were stored at 4°C. This procedure was found to preserve the cells at least for 5 days (data not shown). Preservation of the solution by using bismuth nitrate oxide dissolved in PBS as described below for *Azoarcus* sp. DSM 9506 was also tested, but caused cell aggregation. Cells were detached from the beads using 10 mM tetrasodium pyrophosphate dissolved in the fixation buffer as described for

Azoarcus sp. DSM 9506. After pelleting crushed bead fragments, 2 ml of the aqueous phase was transferred to a glass tube, and the cells were washed and stained for DNA determination as described previously [50]. DNA labelling was optimised for the field samples with regard to DAPI concentration and staining time. The procedure was found to label all cells when using a DAPI concentration of 0.68 μM and a staining time of at least 10 min. This was confirmed microscopically.

3.4 Azoarcus *sp. DSM 9506 as a Test Organism*

Azoarcus sp. strain DSM 9506 was used as a reference strain to establish and verify if flow cytometry can be successfully applied in combination with the in situ microcosm approach. The strain was obtained from the DSMZ (*German Collection of Microorganisms and Cell Cultures, Braunschweig, Germany*). The strain was anaerobically grown in batch cultures with toluene as the sole source of carbon and energy using the mineral salt medium 586 of DSMZ. Tracer solution SL-10 (DSMZ medium 320) and vitamin solution (DSMZ medium 461) were added to the medium. The medium was completed by adding 18.7 ml of 1 M $NaHCO_3$ and 18.7 ml of 1 M KNO_3 l^{-1}. All solutions were sterilised by filtration or autoclaving and flushed with nitrogen before use. The medium was subsequently prepared in an anoxic glove box (gas atmosphere 95% nitrogen, 5% hydrogen; Coy Laboratory Products Inc., Grass Lake, MI) to ensure anoxic conditions. For cultivation, 118 ml serum bottles were each filled with 50 ml of medium and 55 Bio-Sep® beads, which were used as microcosm material. Before use, the beads were heated to 300°C for 4 h. The activated carbon matrix of the beads provides a surface that microorganisms can colonise and that serves as an adsorbent for toluene to prevent toxic effects. Serum bottles were amended each with 26 μl toluene and immediately closed with Teflon-coated butyl rubber stoppers (ESWE Analysentechnik Gera GmbH, Germany). The serum bottles were incubated at 30°C for 24 h before inoculation in order to allow uniform adsorption of the toluene to the bead material. Subsequently, the microcosms were inoculated with 5% of a pre-culture of *Azoarcus* sp. DSM 9506 grown to the late exponential phase and incubated statically at 30°C in the dark.

For single cell analytics, ten beads of each serum bottle (in triplicate) were taken under anoxic conditions after 25, 49, 72, 95, 167 and 213 h of incubation and crushed mechanically with a sterile glass bar. Samples were fixed immediately with 5 ml fixation buffer consisting of 5 mM phosphate buffered saline (PBS), 5 mM bismuth nitrate oxide and 10% sodium azide [11]. For detaching the cells from the matrix, different concentrations of tetrasodium pyrophosphate, Brij 35, Tween 20, sodium ethylene-diamine-tetra-acetate (EDTA), and sodium chloride were tested. The addition of tetrasodium pyrophosphate to the fixation buffer at a final concentration of 10 mM followed by gentle shaking of samples for 45 min was found to result in the highest efficiency of cell detachment with an efficiency of 61%. The sample was centrifuged for 3 min at 246×*g* to pellet the bead material. The detached cells were stored in glass vials at 4°C until performing the DNA staining procedure (see below). The fixation procedure was proven to preserve the cells in a stable state for 3 days (data not shown).

3.5 Cell Enumeration and Analysis of Multiplication of Bio-Sep® Bead-Grown Bacteria

The laboratory studies with *Azoarcus* sp. DSM 9506 were used to specifically establish (1) if it was possible to detach structurally intact cells from the porous Bio-Sep® beads, (2) to determine the sensitivity of the method to detect changes in cell numbers during anaerobic growth, and (3) the resolution of the method by analysing proliferation activity of *Azoarcus* sp., i.e., varying chromosome contents. The microcosms were amended with toluene as carbon and energy source and nitrate as an electron acceptor. Every few days (see above for *Azoarcus* sp. cultivation), ten beads were harvested from the laboratory microcosms and cell numbers, scatter signals, and DNA patterns were analysed (Fig. 1). The intact cell structure was confirmed by strictly clustered cell distributions in a histogram

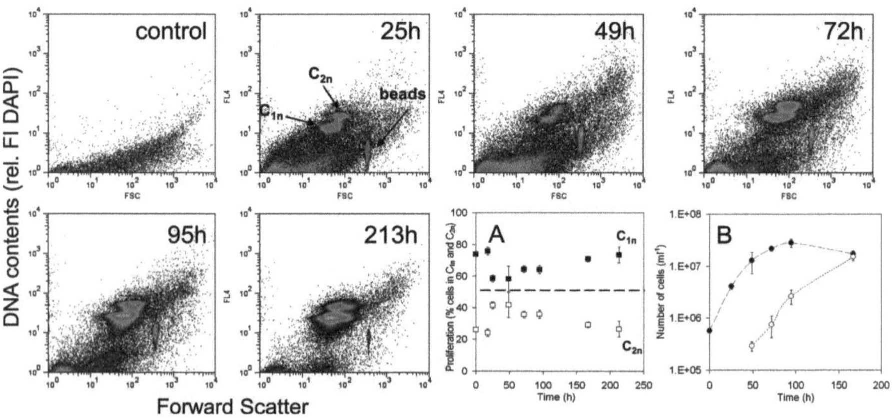

Fig. 1 Growth of *Azoarcus* sp. strain DSM 9506 on toluene as carbon and energy source and nitrate as an electron acceptor. Cultivation was performed using Bio-Sep® beads cultivated in static batches at 30°C. The control was obtained from sterile beads. The beads were crushed and handled as described in Experimental Procedures for flow cytometric investigation. The cells were harvested and labelled with DAPI for DNA quantification and the sub-populations fractions estimated for changing proliferation activity **a**. The bacteria were also analysed together with alignment microspheres (1.0 μm) for cell number determination both in the cultivation broth and within the crashed Bio-Sep® beads **b**. Clearly, increases in cell numbers (closed symbols = cells in solution, open symbols = cells in Bio-Sep beads) and sub-population shifts are in agreement, providing evidence for reliable quantification and activity determination of bacteria grown on surfaces and within pores of Bio-Sep® beads

displaying forward scatter (FSC) and side scatter (SSC, data not shown) features of the cells. Communities dominated by cells with disturbed cell walls and membranes would not show such a discontinuous scatter behaviour. Growth was confirmed by increases in cell numbers up to around 80 h, representing about three duplications. The DNA pattern confirmed this activity. *Azoarcus* sp. DSM 9506 represented two sub-populations with different chromosome contents during anaerobic growth on toluene. The C_{2n} sub-population with the higher DNA content is a marker for growth since the percentage of C_{2n} cells increased in parallel to increases in cell numbers. During the stationary growth phase cells with the single DNA content (C_{1n}) were dominant. All data were obtained by analysing at least three biological replicates. The flow cytometry approach developed and tested here displayed high sensitivity and resolution and was therefore considered to be a suitable protocol to analyse microbial communities colonising in situ microcosms.

3.6 Microscopy

DNA-DAPI-stained cells were subjected to epifluorescence and phase contrast microscopy (Axioskop, Zeiss; camera: DXC-9100P). The Zeiss filter set 02 (excitation G 365, BS 395, emission LP 420) was used for examining blue fluorescence of DAPI. Visualisation of merged phase contrast and fluorescence images was performed using the Openlab 3.1.4 software (Improvision, USA).

3.7 Multiparametric Flow Cytometry

Flow cytometric measurements were carried out using a MoFlo cell sorter (DakoCytomation, Fort Collins, CO) equipped with two water-cooled argon-ion lasers (Innova 90C and Innova 70C from Coherent, Santa Clara, CA). Excitation at 488 nm (with 400 mW) was used to analyse the forward scatter (FSC) and side scatter (SSC) as trigger signal at the first observation point, using a neutral density filter with an optical density of 2.3. DAPI was excited ML-UV (333–365 nm, 100 mW) at the second observation point. The orthogonal scatter signal was recorded after reflection by a 555 nm long-pass dichroic mirror, passage by a 505 nm short-pass dichroic mirror and a BP 488/10. The orthogonal DAPI signal was passed through a 450/65 band pass filter. Photomultiplier tubes were obtained from Hamamatsu Photonics (models R 928 and R 3896; Hamamatsu City). Amplification was carried out at linear or logarithmic scales, depending on the application. Fluorescent beads (Polybead Microspheres: diameter, 0.483 μm; flow check BB/Green compensation Kit, Blue Alignment Grade, ref. 23520, Polyscience, USA) were used to align the MoFlo (coefficient of variation, CV value about 2%). Also, an internal DAPI-stained bacterial cell standard was introduced for tuning the device up to a CV value not higher than 6%.

Cell numbers were counted accurately (and with negligible deviation) in the cell suspensions using flow cytometry. Fluorescent beads (from a 5 μl stock solution of Fluosheres polystyrene microsperes 1.0 μm (505/515), Molecular Probes; as above) were mixed with the DAPI-stained cells. The dot plots were gated with regard to the sub-communities and the beads, and cell counts were calculated. Cell aggregation was not observed; thus, clearly separated sub-populations were found.

Proliferation activity was determined by evaluating the flow cytometric histograms obtained using both FlowJo (Tristar, Switzerland) and WinList 2.01 (Verity Software House, Maine, USA) software. The dot plots were gated in a way that visibly distinct sub-communities were included in the gates and cell counts therein were calculated. The percentage of cells of any sub-community with a distinct DNA content was counted. Forward scatter (FSC) signals gave additional information.

Cell sorting was performed using the four-way sort option at high speed (12 m s^{-1}). The most accurate sort mode (single and one drop mode: highest purity 99%) was chosen for separating 5,000 cells per second. The cells were sorted into nucleic acid free glass flasks. Cells were separated from the whole community using DNA-DAPI fluorescence intensity and forward scatter signals in several independent experiments using different gate settings. Dominant and apparently growing sub-communities were separated in order to facilitate their molecular identification. Up to four sub-communities per sample could be separated simultaneously. Between 10^4 and 10^5 cells per sub-community, depending on the abundance of cells within the sub-community, were sorted for further phylogenetic analysis.

3.8 DNA Preparation, Cloning and Sequencing of Bacterial 16S rRNA Genes

DNA for whole community analysis was extracted directly from the beads according to Maher et al. [27]. DNA from sorted sub-communities was prepared as described by Günther et al. [12]. For generating 16S rRNA gene clone libraries, bacterial 16S rRNA gene fragments were PCR-amplified using the bacteria-specific primers 27F and 1492R [20] and cloned as described elsewhere [19]. From each of the microcosm variants, 96 clones were collected and screened by restriction fragment polymorphism analysis. Based on the restriction patterns, representative clones were partially sequenced using the sequencing primers 27F and 519R [20]. A total of 130 clones representing 22 operational taxonomic units (OTUs) were selected for partial sequencing. Sequence analysis was done as described previously [31]. The determined partial 16S rRNA gene sequences were deposited in the GenBank database under accession numbers HM217219-217348.

3.9 T-RFLP Profiling of Bacterial 16S rRNA Genes

T-RFLP analysis of 16S rRNA amplicons was performed according to Günther et al. [12] with the restriction endonucleases *Alu*I or *Bst*UI. Relative peak areas were determined by dividing the individual T-RF area by the total area of peaks within the range of 50 to 500 bp. Theoretical T-RF values of the representative phylotypes represented in the clone library were calculated using the NEB cutter (http://tools.neb.com/NEBcutter2) and confirmed experimentally by T-RFLP analysis using the corresponding clones as templates (Table 2). Relative T-RF abundances of representative phylotypes were determined based on the relative peak areas of the corresponding T-RF.

Table 2 Sequencing results of representative 16S rRNA gene clones and experimentally determined terminal restriction fragments (T-RF) of the respective operational taxonomic units (OTU)

OTU	Acc. no.	Taxonomic affiliation according to RDP 10	T-RF *Alu*I (bp)	T-RF *Bst*UI (bp)
1	HM217219-217269	*Achromobacter* sp.	149	141
2	HM217270-217279	*Ralstonia* sp.	69	390
3	HM217280	*Burkholderia* sp.	228	381
4	HM217281-217284	*Ochrobactrum* sp.	202	88
5	HM217285	*Acidovorax* sp.	145	201
6	HM217286	Comamonadaceae	247	173
7	HM217287	*Magnetospirillum* sp.	205	195
8	HM217288	*Azoarcus* sp.	147	139
9	HM217289	Rhodocyclaceae	232	115
10	HM217290-217299	*Geobacter* sp.	67	105
11	HM217300-217306	*Geobacter* sp.	249	105
12	HM217307	*Geobacter* sp.	149	141
13	HM217308	Peptococcaceae 1	71	234
14	HM217309	Peptococcaceae 1	312	174
15	HM217310-217223	Peptococcaceae 1	222	110
16	HM217224-217331	Peptococcaceae 1	271	161
17	HM217332-217337	Peptococcaceae 1	85	124
18	HM217338-217343	*Acetivibrio* sp.	232	32
19	HM217344-217345	*Acidaminobacter* sp.	236	123
20	HM217346	*Prolixibacter* sp.	184	99
21	HM217347	TM7 phylum	435	58
22	HM217348	Unclassified Bacteria	226	393

3.10 Statistical Analyses

3.10.1 Analysis of T-RFLP Profiles

For estimating differences in whole community composition, T-RFLP data were analysed by non-metric multidimensional scaling (n-MDS) based on Bray-Curtis similarity and minimum of 40 repetitions. Prior to the statistical analysis, T-RFLP signals were separated from noise following the statistical approach provided by Abdo et al. [1] using a standard deviation of 3 as cutoff level. The resulting normalised data were used as the basis for the following n-MDS analysis. All analyses were carried out with the statistical computational environment R Version 2.10.1 (R Development Core Team 2009) and the R package 'vegan' Version 1.17-1 [36].

3.10.2 Analysis of Flow Cytometry Data

To identify variations in cluster distributions in flow cytometry bivariate plots, a newly developed combination of image analyses with a multivariate approach was used. This approach was called the Dalmatian plot tool. Similarities of the single plots were estimated by direct comparison of the various plots to each other and the subsequent estimation of the rate of overlap between the occurring clusters indicated by sort gates (Fig. 2). Two different approaches were tested. For both approaches the gates were extracted from the flow-cytometric bivariate plots using FlowJo (Tristar, Switzerland) and transferred into bitmap pictures of 390 × 390 pixels in size. In the first approach, only presence-absence of clusters was regarded, enabling equal priority of all emerging clusters independently of the abundance of cells within. For the analysis, binary pictures were produced with gates filled with black color and a white background (Fig. 2). By simple image calculation, overlays of all picture combinations were produced. For estimating overlap rates, the area sums from all gates of the original binary pictures were calculated as well as for the overlaps. Similarities were then estimated using a modified Jaccard Index S given by:

$$S_{A_1A_2} = \frac{\sum (A_1 + A_2) - 2 \times \sum A_{\text{total}}}{\sum (A_1 + A_2)}$$

with A_1 and A_2 as the sum of all gate areas in pixels in each of the two pictures, and A_{total} the sum of gate areas of the overlap of two pictures. Overlay creation and gate area calculation were automatically done with ImageJ Version 1.43 (http://rsb.info.nih.gov/ij). The similarity results of all possible combinations were then transferred into a triangular similarity matrix. The final n-MDS analysis based on the similarity matrix was accomplished under R as described above.

For the second and slightly different approach, cell abundances of the gates were included for calculating similarities. Gated clusters were again extracted and transformed into 8-bit pictures providing 256 (= 2^8) grey color steps, and relative

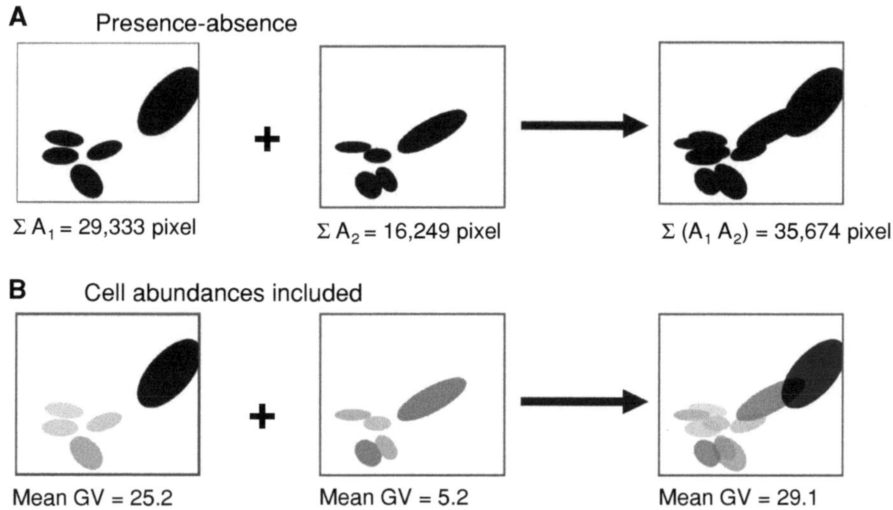

Fig. 2 Illustration of methodology used for estimating community similarities of cytometric flow plots using a Dalmatian plot. Areas of gates were estimated as **a** sum of pixels for presence–absence and **b** grey values when cell abundances were taken into account. Sums were calculated from plots of each of the samples separately and for the overlap of two samples, respectively. For similarity estimation a modified Jaccard index was used (for more details see the Experimental Procedures section)

cell numbers of gates were then translated into grey values (Fig. 2b). To set the correct grey value numbers, all gate cell numbers were calculated as relative to the occurrence of the maximum cell number found in one gate of all pictures. Cell numbers were then set to relative grey value of 255 (= black) for maximum occurrence and 0 (= white) for areas where no cells (= outside of gates) were present. Finally, all calculated grey values were divided by 2. The latter step was necessary to prevent occasions in which two overlapping areas may together result in grey values >255. Instead of estimating gate areas, we now calculated the mean grey value (mean GV) for each picture and their resulting overlaps as proxy for calculating similarities. Mean GV is estimated as the sum of all grey values occurring in a picture divided by its pixel number (which was a constant of $390 \times 390 = 152{,}100$ pixels in our case). Mean GVs from each picture and all overlay combinations were then used as a basis for calculating similarities and subsequent n-MDS analysis as described above.

4 Variations in Composition and Structure of Natural Microbial Communities

The goal of our study was to evaluate a combined approach of community fingerprinting, flow cytometry, and trend interpretation analysis to study the effect of various electron acceptor conditions on potentially active members of the indigenous microbial community in a BTEX contaminated aquifer. Five microcosm variants (see Exposure of in Situ Microcosms) were each exposed in three monitoring wells differing in their hydro-geochemical characteristics. Alterations in community structure and composition influenced by hydro-geochemical characteristics (present in the different wells) and micro-environmental conditions (kind of electron acceptor added) were studied using both molecular tools and single cell analytics. For the latter, cells were quantitatively associated to clearly defined clusters of sub-communities that alter in dependence on well conditions and supplied electron acceptors. Whereas flow cytometry was used and defined to follow alterations in community structure, the phylogenetic affiliation of abundant community members was performed to obtain insights into variations in community composition.

4.1 Similarity of Community Composition According to T-RFLP Patterns

Total DNA extracted directly from each microcosm was subjected to T-RFLP analysis of amplified bacterial 16S rRNA genes to assess shifts in the whole microbial community composition. Whether the hydro-geochemical conditions or the supplied electron acceptors influenced the community composition was estimated by n-MDS analysis (Fig. 3). Symbols in the plot located closer together indicate more similar community composition, while those placed more distantly are more dissimilar.

The n-MDS plot of T-RFLP data clearly highlighted the major influence of the hydro-geochemical conditions on community composition (Fig. 3). Significant differences in community composition with regard to phylogenetic identity and relative abundances of the respective abundant phylotypes can be stated (Fig. 4). The respective T-RFLP patterns of the microbial communities retrieved from each of the three wells were grouped closely. This held particularly true for well 7/99.

In addition to the influence of hydro-geochemical parameters in different wells, the microcosm treatments also had an influence on community composition. Especially the microcosms loaded with nitrate were separated from all other treatments, indicating distinct community compositions due to the supplied nitrate. Sulphate-amended microcosm communities from the different wells were also clearly distinct. The T-RFLP profiles of all untreated microcosms (blanks) showed smaller differences to each other, indicating the influence of the hydro-geochemical conditions to be potentiated by the electron acceptor and benzene amendment.

The distances shown in the n-MDS plot (Fig. 3) are based on the phylogenetic composition of every microcosm community (Fig. 4). In highly benzene-contaminated wells (7/99, 18/00), the dominant phylotype grown in all

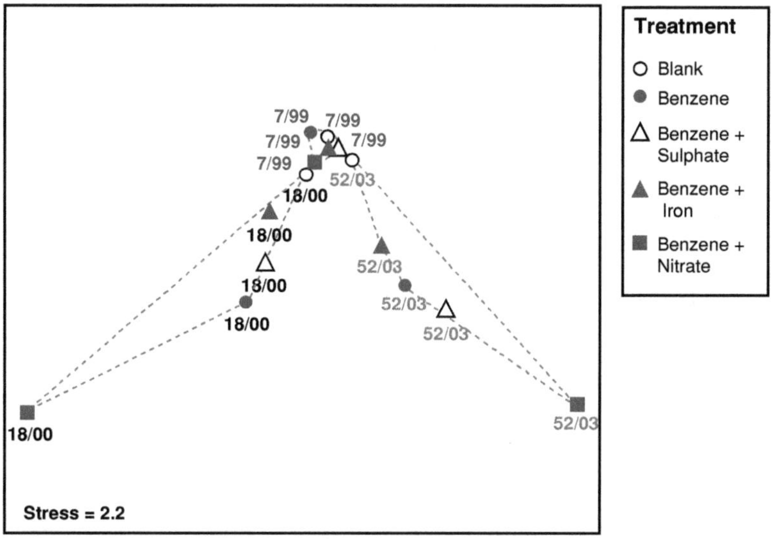

Fig. 3 N-MDS analysis plot of T-RFLP patterns of wells 7/99, 18/00 and 52/03 from microcosms amended with benzene, benzene + sulphate, benzene + iron and benzene + nitrate, respectively. Blank (unamended) microcosms served as references. N-MDS analysis is based on the Bray-Curtis index

microcosms but the nitrate-amended ones was affiliated with *Achromobacter* spp. (relative T-RF abundances of 56–92%; Fig. 4a, b). Other phylotypes found in most of the microcosms belonged to *Ralstonia* sp. (3–10%) and *Burkholderia* sp. (2–3%). Pronounced shifts between the two groundwater wells were observed in the remaining T-RF. Phylotypes affiliated to *Acidovorax* sp. were found in all benzene-amended microcosms within well 18/00, with relative T-RF abundances of 5–16%. Phylotypes affiliated to *Geobacter* sp. were found in most of the benzene-amended microcosms within both wells; however, the relative abundance of this T-RF increased dramatically in the nitrate-amended microcosm within well 18/00, indicating a stimulation of *Geobacter* sp. by nitrate supply in well 18/00, but not in 7/99.

The T-RFLP profiles generated from well 52/03 revealed a different community development compared to the other two wells (Fig. 3, 4). This can be attributed to the fact that the well is located at the fringe of the contamination plume and therefore has distinct hydro-geochemical conditions. The n-MDS plot (Fig. 3) strengthens these findings for this community. Generally, a higher diversity of the microbial community was observed (Fig. 4c). Additionally to *Achromobacter* sp. (in this well abundant with 22–85%), another abundant phylotype belonging to the *Peptococcaceae* group 1 (with three OTUs according to the RDP taxonomy, 12–35% in total, not in the blank microcosm) was detected. Further phylotypes that were detected only in well 52/03 belonged to the *Rhodocyclaceae* (3–4% in all treatments except for blank and nitrate-amended microcosms), to the genus

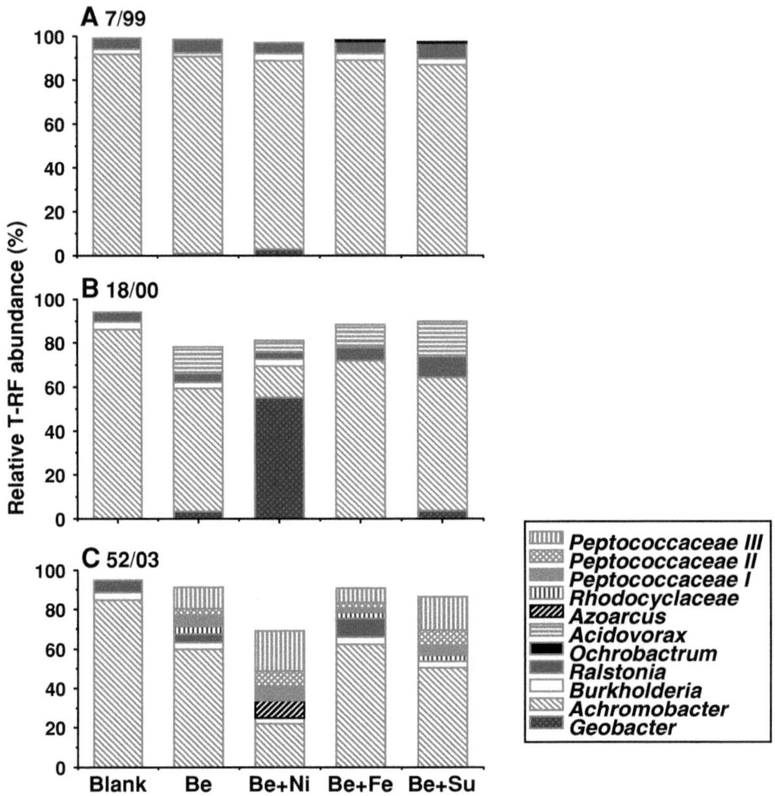

Fig. 4 Phylogenetic composition of the communities harvested from the microcosms exposed in well **a** 7/99, **b** 18/00 and **c** 52/03. T-RFLP profiles were generated using the restriction endonuclease *Bst*UI. *Be* benzene; *Ni* nitrate; *Fe* iron; *Su* sulphate

Azoarcus (9%, only in the nitrate-amended microcosm) or *Acetivibrio* sp. (only with nitrate or sulphate amendment, compare also clone library, Table 2). Similar to the other two wells, phylotypes belonging to *Burkholderia* sp. (3–4% in all treatments) and *Ralstonia* sp. (4–9% in all treatments except for nitrate and sulphate) were detected, but obviously played a minor role. To summarise the data from well 52/03, microcosms loaded with benzene alone and with benzene and iron showed similar community compositions. Microcosms amended with nitrate and sulphate varied in terms of different relative abundances of all phylotypes detected in the respective well and by the presence of treatment-specific phylotypes, such as *Azoarcus* sp. or *Acetivibrio* sp.

Recapitulating, the T-RFLP profiles showed that the hydro-geochemical conditions shaped the microbial community composition, which was, however, also influenced by the specific microcosm treatments. The community compositions in blank microcosms were all found to be similar due to the hydro-geochemical environment. Apart from the grouping of the data according to the respective well, nitrate-amended microcosm communities were the most distinct, followed by communities in sulphate- and benzene-amended microcosms.

4.2 Similarity of Community Structure as Inferred from DNA Patterns

Apart from comparing community compositions based on T-RFLP data, it is also possible to compare community structures based on DNA versus scatter patterns. This approach relies on single cell analytics and the use of the fluorescent dye DAPI, which specifically binds to AT-rich regions of the DNA. Bacteria contain single chromosomes or several copies depending on the growth state. The light that is refracted by the cells gives additional typical cell-based information. Both parameters lead to fingerprint-like cytometric patterns of subsets of cells if the bacteria are analysed one by one with regard to DNA contents and light scatter characteristics using a flow cytometer. If several species are present, cells will gather in numerous clusters with changing cell abundances therein. The position of such clusters and the respective cell abundances give further information on community stability over a time range. Community structure may change according to micro-environmental conditions as was investigated here.

Flow-cytometric community pattern analysis was compared to T-RFLP community data. Typical pattern distributions are shown in Fig. 5 for the microcosms of well 52/03. The upper three histograms show very similar distributions. Although the benzene concentration is low in well 52/03, the basic community pattern of the blank sample resembles that of the benzene-amended microcosm of well 52/03. For comparison, the histogram of the benzene-treated microcosm of the highly benzene-contaminated well 18/00 is also shown, and even here a similar pattern is visible. The DNA patterns obtained from communities isolated from iron- and sulphate-amended microcosms were similar to each other, whereas the nitrate-amended community displayed a DNA pattern that was dissimilar to those of all other microcosm variants.

As before, n-MDS plots were used to interpret the complex data. Subsets of cells were virtually gated, and the number of cells inside the gates was determined using the program FlowJo. The position of the gates in a histogram created the basis of the Dalmatian-like pattern, named after the shape and colour of the gates also invented in the n-MDS plot analysis. Data from all microcosms were analysed in the same way and the complete number of events set to 100% with the exception of the noise cluster within the lower left corner of the histograms. The positions of the gates were rated via n-MDS analysis (Fig. 6). Additionally, cell numbers per gate were used to obtain a higher sensitivity of the similarity analysis (Fig. 7). It was surprising to see that either-or decisions like using only the gates' positions as was done with the Dalmatian plots clearly resulted in a contrasting predication compared to the T-RFLP community data sets (Fig. 3). According to DNA pattern

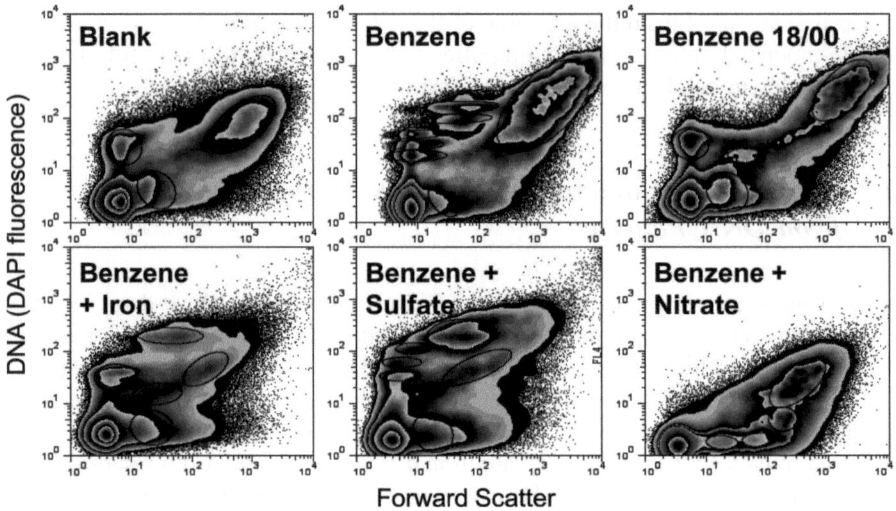

Fig. 5 DNA/FSC pattern distributions of cells harvested from the microcosms exposed in well 52/03. For comparison, the distribution of the benzene-treated microcosm harvested from well 18/00 is also shown. Up to 250,000 cells were analysed and the dominant sub-populations were gated. The peak in the lower *left corner* of the histograms represents the noise of the cytometer as well as unstained cell debris and was therefore not gated. All other upcoming sub-communities are indicated by gates

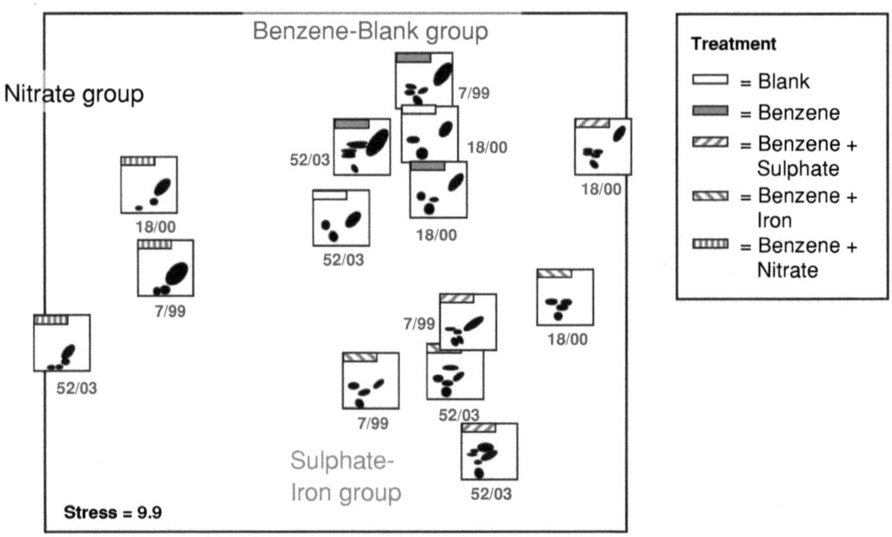

Fig. 6 Dalmatian-n-MDS analysis with overlaid flow plot results derived from microbial microcosm communities using presence-absence of community gates based on flow cytometry analysis. *Black patches* in gate pictures depict gate positions. Sample treatments are indexed in *square* given in the *upper left corner* of each flow plot picture

distributions, the treatment of the microcosms and not the hydro-geochemical conditions determined the similarities of the developing communities. Three distinct clusters were grouped, one comprising all nitrate-amended microcosms, the second one comprising sulphate- and iron-amended microcosms, and a third one in which the highly similar communities from microcosms with only benzene amendment and from those without any treatment were grouped. This scenario suggests a strong dependence of the community structure on the microcosm treatment. It also shows that the application of benzene to a microcosm in the fringe well (52/03) shifted the community structure so that it became similar to that of the highly contaminated wells.

Using only the position of gates for an n-MDS analysis is a strict approach (Experimental Procedures, Fig. 2), but allows all gates to be judged independently of the abundance of the bacteria therein. This might be of advantage given that often highly abundant bacteria in a distinct cluster are not necessarily those most relevant for the studied biogeochemical process, but represent functional generalists. Such generalists might tolerate a wide range of environmental conditions and be prominent in the majority of samples. The perceived importance of such widespread bacteria can be down-weighted, while less abundant but potentially important bacteria can be amplified by levelling species abundance to presence–absence decisions.

However, it may as well be that high cell numbers reflect high bacterial activity. In an alternative data treatment, cell numbers in gates were therefore included in

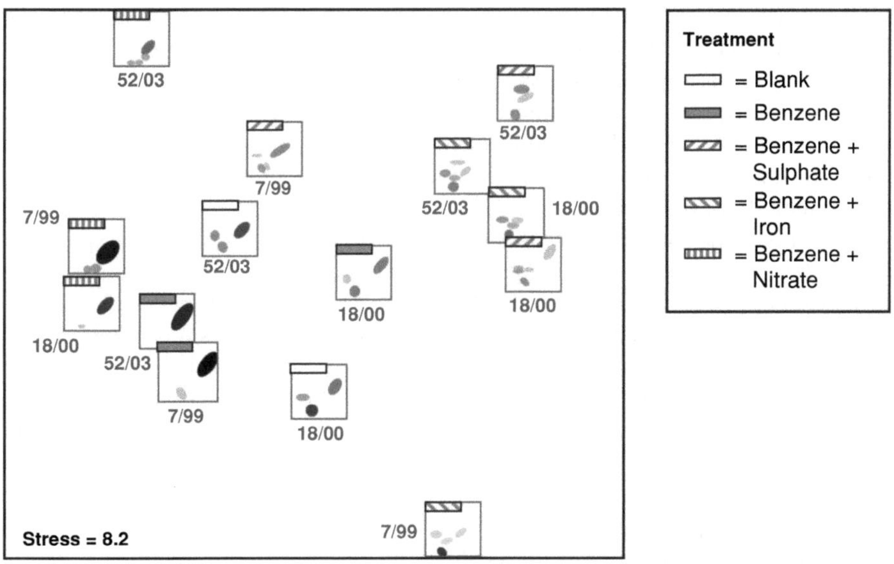

Fig. 7 Dalmatian-n-MDS analyses with overlaid flow plot results derived from microbial microcosm communities including cell abundances within gates. Cell abundances are given as relative grey value within gates. Sample treatments are indexed in the *square* given in the upper left corner of each flow plot picture

the comparison (Fig. 7). It seems indeed reasonable to include cell numbers into the n-MDS approach since they varied largely. In the best case, cell numbers may correlate with BTEX degradation capacities or other important microbial activities related to the microcosm amendment. As before, where cell abundances were disregarded (Fig. 6), the resulting n-MDS plot (Fig. 7) indicated a stronger influence of the microcosm treatment relative to the hydro-geochemical conditions. However, when cell numbers were included, the trends were less pronounced. Principally, when including cell numbers in the n-MDS analysis, a gradient of changing species dominance resulted. Microcosm communities with one dominant microbial group were positioned in the lower left part of the plot, while those with levelled species dominance were located in the upper right corner. Again, nitrate-amended communities stood out on the left side of the plot. The blank and benzene-only-amended microcosm communities grouped together, this time in the middle of the plot. Only the sulphate- and iron-amended microcosm-derived data were more widely distributed than in Fig. 6. Again, there was no obvious correlation with the hydro-geochemical conditions. Samples 52/03 nitrate and 7/99 iron, however, were distinct because these samples contained, due to limited cell detachment from the bead material, only extremely few cells.

4.3 Identification of Key Players in Sorted Sub-Communities

Statistical analysis of T-RFLP profiles and cytometric DNA pattern analysis can provide information on changes and trends in microbial community structure resulting from varying micro-environmental conditions. The limitations of each single approach, namely the limited sensitivity and quantitative resolution of the whole community by T-RFLP profiles and the lack of phylogenetic information in pure flow cytometric analyses, can be overcome by combining both approaches and by including cell sorting.

To demonstrate the benefit of this combined approach, microcosms exposed in well 52/03 were selected for detailed analysis of selected sub-communities using flow cytometry, cell sorting, and T-RFLP analysis. As shown in Fig. 5, specific sub-communities emerged in the DNA versus FSC scatter plots depending on the treatment of a chosen microcosm. As an example, one specific sub-community that was visible in all but the blank and the nitrate-amended microcosms was selected for cell sorting. Sort gates were placed around the sub-communities as indicated in Fig. 8 (gate 2), and cells within the gates were physically separated by cell sorting for DNA extraction and T-RFLP profiling of bacterial 16S rRNA genes. The phylogenetic composition of the two selected sub-communities in comparison to the whole community composition is shown in Fig. 8. In both treatments, gate 2 comprised mainly the *Peptococcaceae* phylotype that was also present in both whole communities, but in significantly lower proportions. By contrast, the other phylotypes dominating the whole communities, most notably *Achromobacter* sp., were not detected in the sorted sub-communities.

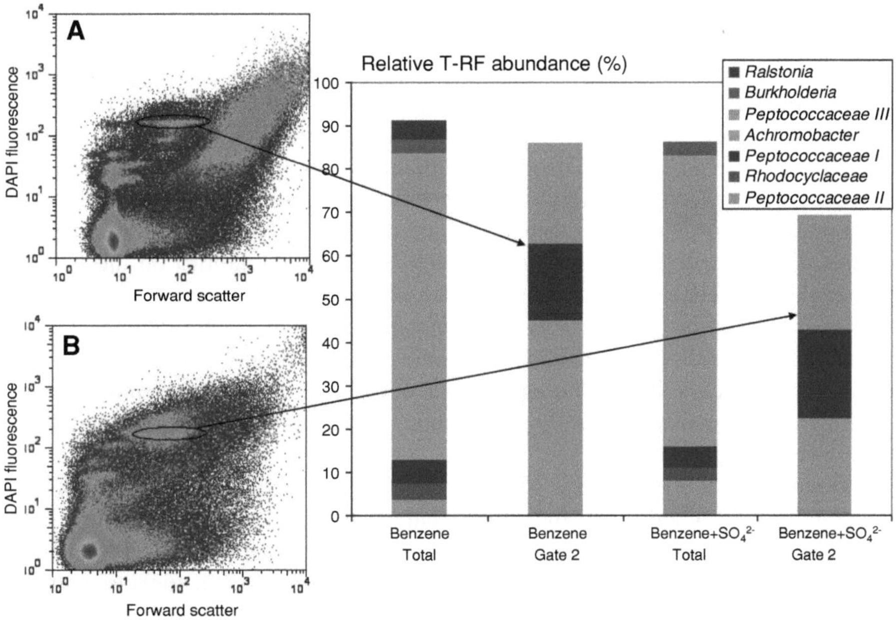

Fig. 8 Patterns of DNA distribution (DAPI fluorescence) versus FSC signal of cells harvested from microcosms exposed in well 52/03 (*left panels*). **a** Microcosm treated with benzene. **b** Microcosm treated with benzene and sulphate. *Right panel* Community composition of whole communities and sorted sub-communities of gate 2 according to T-RFLP profiling with the restriction endonuclease *Bst*UI

5 Significance of the Combined Approach for Ecological Interpretation and Implications for Biostimulation

The combined use of T-RFLP profiling, cytometric DNA community pattern analysis, and n-MDS similarity analyses provided insights into variations in microbial community compositions and structures within in situ microcosms. The newly developed Dalmatian plot evaluation of the cytometric data made an important contribution for understanding changes in the community dynamics of the microcosms. Our results indicate that T-RFLP profiling and the Dalamtian plot evaluation of the cytometric data contributed to the same extent to the interpretation of the microcosm community developments. As an outcome of the combined approach, we found that hydro-geochemical conditions as well as microcosm treatments with various electron acceptors affected the community composition and structure. Such information increases the understanding of the contaminated ecosystems and might contribute to the development and control of future remediation strategies.

In this study the cytometric pattern analysis indicated greater effects by electron acceptor amendments in all three wells than similar analyses of T-RFLP profiles.

The same treatments performed in the different wells resulted in similar community structures. These peculiar findings were substantiated when cell numbers were also accounted for cytometric n-MDS analysis (Fig. 7). At first glance the results obtained by T-RFLP and cytometric pattern analyses seem to be contradictory. However, this can be explained by the different targets of both approaches: T-RFLP analysis targeted whole communities on the basis of the nucleic acid sequence and reflects only those members that are present to a certain levelled amount and whose sequence can be assorted using specific restriction endonucleases. Cytometric pattern analysis targeted whole communities on the level of the single cell. These cells cluster into subsets where cells contain stained chromosomes of similar size and number and have similar light scatter refraction features. As a result, specific sub-communities emerge from the background of the overall community. Thus, both approaches complement each other and thus provide a comprehensive insight into indigenous community composition and structure development.

To summarise, most of the phylotypes found are likely to degrade benzene under microoxic conditions, indicating that a specific benzene-degrading community colonised the microcosms. The co-occurrence of aerobic, nitrate-reducing, iron-reducing and possibly fermentative bacteria indicates the development of ecological micro-niches in the in situ microcosms. Here, the hydro-geochemical parameters set the master conditions under which biostimulation affected community composition to a lower and community structure to a higher degree.

5.1 Benefit of the Phylogenetic Data Sets

The more specific outcomes are discussed in the following. In wells 7/99 and 18/00, the hydro-geochemical conditions were characterised by high benzene concentrations and low dissolved oxygen concentrations (below 0.09 mg l^{-1}) in addition to low redox potentials, indicating the dominance of anaerobic metabolisms in this part of the plume. Sulphate and nitrate with concentrations lower than 10 mg l^{-1} were strongly depleted compared to the pristine aquifer (around 1,000 mg l^{-1} for sulphate and 275 mg l^{-1} for nitrate), suggesting sulphate reduction and nitrate reduction as dominant terminal electron acceptor processes. The main difference between these two wells was the higher iron concentration in well 18/00. As dissolved iron concentrations arise mainly from ferrous iron, iron reduction seems to be an important process in well 18/00. However, aerobic and facultative anaerobic nitrate-reducing bacteria were the dominant species involved in the colonisation of the microcosms. The microbial communities were dominated by a phylotype affiliated to the genus *Achromobacter*, a genus that is known to comprise aerobic BTEX degraders [35]. The dominance of this genus indicated that oxygen may play an essential role for biodegradation of benzene in those wells. The other two phylotypes found in the most of the microcosms were affiliated to the genera *Burkholderia* and *Ralstonia*. These organisms are known for their metabolic versatility, and several species degrade aromatic compounds under both oxic and micro-oxic conditions [16]. The enrichment of these bacteria in the benzene-amended microcosms points additionally at benzene degradation with oxygen as electron acceptor. This dominance of the aerobic and nitrate-reducing bacteria might be explained by a rapid consumption of the incoming oxygen and nitrate, enabling the occurrence of BTEX degradation by both aerobic as well as nitrate-reducing processes.

Well 52/03, located at the fringe of the contamination plume, was characterised by low benzene concentrations and relatively high nitrate and sulphate concentrations compared to the other wells. Although oxygen concentrations were below 0.06 mg l^{-1}, oxygen seemed also to play an important role in biodegradation processes in this well as indicated by the dominance of the already-mentioned phylotypes of *Achromobacter*, *Burkholderia* and *Ralstonia*. However, the different hydro-geochemical conditions of this well compared to the others effected an altered composition and increased phylogenetic diversity of its community when benzene and the various electron acceptors were supplied (Figs. 3, 6, 7). Dissimilar from the highly contaminated wells, phylotypes of *Azoarcus* and *Rhodocyclaceae* also appeared. The *Rhodocyclaceae* comprise genera such as *Azoarcus* and *Thauera*, which harbour several species known for BTEX degradation under nitrate-reducing conditions (for a review see [43]). Members of the *Peptococcaceae* were also detected in well 52/03. This family, belonging to the *Clostridia*, comprises gram-positive, strictly anaerobic, spore-forming bacteria, which are known for a variety of biochemical activities, e.g., fermentation as well as sulphate, iron and nitrate reduction. Therefore, it is difficult to judge the specific role of these organisms here. It is possible that they grow within anaerobic micro-niches generated by oxygen consumption of fast-growing benzene degraders and use metabolites or dead biomass accumulating within the colonised microcosms after the added benzene has been consumed. Interestingly, they were absent in the blank microcosm.

When evaluating the changes caused by the diverse electron acceptor treatments, it was found that nitrate amendment had a strong influence on the community compositions in wells 18/00 and 52/03 according to the T-RFLP profiles. Because well 18/00 is located in the highly contaminated zone where no nitrate was detectable, the supply of nitrate obviously has stimulated nitrate-reducing BTEX degraders here. In addition to the already above-described aerobic genera *Achromobacter*, *Ralstonia* and *Burkholderia*, which belong to the Burkholderiales of the Betaproteobacteria, an aerobic *Acidovorax* species was found in relatively high proportions in this well. Although all these genera are expected to degrade benzene using oxygenases and thus may depend on molecular oxygen, nitrate respiration can also be widespread among them, but has still not been described as an electron-accepting process for benzene degradation by the members of Burkholderiales or *Acidovorax*. Also, phylotypes affiliated to *Geobacter* sp. were found to constitute a very large proportion (nearly 60%) in the nitrate-amended microcosm in well 18/00. Members of the family *Geobacteraceae* are usually iron reducers in subsurface environments [23], but *Geobacter metallireducens*, for example, can also degrade toluene while using nitrate as an electron acceptor [24]. Therefore, aerobic and facultative anaerobic nitrate-reducing bacteria were the dominant species involved in the colonisation of the nitrate-amended microcosms in well 18/00. Significant changes in the microbial community composition were also observed in wells 18/00 and 52/03 when sulphate was supplied. Sulphate concentrations were low in well 18/00, but high in well 52/03. However, sulphate reducers were not detected in any of the investigated wells with the exception of the phylotypes affiliated to *Peptococcaceae*, the ecological role of which is unclear at present. The sulphate amendment obviously did not stimulate sulphate reducers both in the presence or absence of high benzene concentrations when communities were analysed by T-RFLP. The same is true for iron amendment. Remarkably, microcosms deployed in well 7/99, in which potential electron acceptors were strongly depleted, showed no response of the microbial community composition when amended with benzene and any electron acceptor compared to the blank (Figs. 3, 4).

To summarise, the observed changes of microbial community compositions during enhanced bioremediation are in agreement with previous observations made in biostimulation studies [42]. Nevertheless, the addition of electron acceptors to a field site does not guarantee biostimulation because for this approach to be effective the bacteria that are capable of utilising the amended electron acceptor must be present in the first place [7]. Therefore, new technologies are required to assess the effect of various electron acceptors on the microbial community and on biodegradation processes under in situ conditions. In this study the influence of supplied electron acceptors on indigenous community composition was investigated in combination with quantification of emerging potential degraders, which is a novel and effective approach that has not been applied so far.

5.2 Benefit of the Dalmatian Plot Tool

Two-dimensional histograms of cytometric analyses are often complex and contain various kinds of information (e.g., characteristics of events, number of gates, gate positions, cell number within gates), which makes direct comparison often difficult or even impossible, especially when many plots have to be compared. Multivariate statistics like n-MDS, however, enable encompassing multiple simultaneous variables, allow to simplify complex dimensionality and to make complex correlations apparent. For this, the multivariate information of an input matrix is normally broken down to an index number representing how identical two measured objects are by direct comparison of every possible binary combination. Since n-MDS analyses have not yet been applied to cytometric analyses of microbial communities, a close description and discussion of this new approach are provided in this study.

Here, the characteristic cell parameters measured were the forward scatter behaviour of the microbial cells, which is often correlated to cell size, and the

DNA contents made detectable by DAPI staining. Upcoming dominant sub-communities clustered according to these parameters. Virtual gates were set to determine cell numbers of the dominant sub-communities. As a result, dot plots are constructed, which in this study are called Dalmatian plots. For n-MDS analyses the number of gates, their individual positions in the histogram and their form was used as basic information on community structure. Although plots are reduced to gate pictures, the inherent information of the community structure contained is kept within these pictures. Gate areas of these binary, black-and-white pictures were estimated as simply the sum values of all pixels that had the value of 1 (= "black") (Fig. 2a). Community similarities from gates were then estimated by the degree of gate overlaps between all samples. Estimating gate overlaps was used as a suitable proxy of how equal two communities are and moreover allows elucidating the rate of change that has occurred between these two communities. Estimating the degree of overlap directly allows estimating similarities between samples at the presence-absence level.

In a second step cell abundances of gates were included since they varied strongly between gates and should therefore have an influence on the significance of a gate. Given the above-mentioned (see Experimental Procedures) modified Jaccard index, one could down-weight gates relative to the cell number they encompass by multiplication of the individual gate areas with their relative cell numbers. But here initially a difficulty arose since it was cumbersome to estimate the relative amount of cells contained in the overlapping areas and is thus not feasible to estimate their significance directly. The predicament could be circumnavigated by using grey values representing relative cell numbers within gates. Mathematical addition of the pixel grey values of two plots results not only in a new picture containing the overlapping gates of both plots, but also enables an estimate of the relative number of cells in the overlapping areas (Fig. 2). Grey values of overlapping areas result from the relative amount of the cell contents of the respective gates and thus allow to correctly 'weigh' the area where two gates overlap. This simple graphic approach was used for estimating similarities, but now involving cell numbers via grey value estimation (Fig. 2b). Thus, the 'greyness' (the mean sum of all grey values) of a picture represents (1) the total amount of all gates, (2) their overall expansion and (3) all cells that are encompassed by all gates.

Both approaches (presence-absence and inclusion of cell numbers) should be equally used and depend on the underlying hypothesis one assumes. The highest cell abundances in gates may represent those microbial sub-communities that, under the given environmental conditions or experimental treatment, are the key players and therefore the most abundant ones. However, often the most abundant ones are not the key players, but represent species that are most tolerant towards a broad range of environmental parameters. Such euryoec species would then dominate the classification by the n-MDS analyses irrespectively of the samples' origin or treatment. Moreover, it is often not those phylotypes that are most relevant for a given experiment that appear first, but those that grow most rapidly. Usually such so-called r-strategists may then become out-competed by slower growing, but better adapted key players. However, in natural communities there is often a disequilibrium between species abundances. Reasons for dominances in those communities are often complex, depending on a multitude of factors, and not yet completely understood. Thus, putting too much weight on abundances in such situations would presumably lead to only small observable differences between samples since the outcome may be completely determined by only the most abundant species. In statistical analyses of, e.g., plant communities, this problem is normally circumvented by decreasing the differences between species abundances and down-weighting highly abundant species and increasing importance of low abundance species by data transformation (typically square root, double-square root, logarithmic transformation, etc.) prior to analysis [6]. In this study, the same effect was obtained by introducing the grey values for cell numbers in gates. The disadvantage of this approach is that organisms that are of low abundance but highly productive may have no impact on the evaluation of community structure and therefore community functioning.

To summarise, following the row of transformations, the presence-absence (black and white) approach is the most rigorous transformation procedure possible. Its advantage is that its outcome is not dependent on the dominance of bacteria in gates and, moreover, allows clearer estimation of general changes in community structure by easy determination of appearance or disappearance of species (see Fig. 6). Its disadvantage is that equality of gate evaluation may miss the degree of contribution of the cells within the gate to community functioning.

5.3 Benefit of Sub-Community Resolution Down to the Individual Level for Functional Information

Identifying microorganisms responsible for recognised environmental processes has been described as a great challenge in microbial ecology. During the last decade, methodological innovations provided ways of linking the phylogeny and function of uncultivated bacteria via, e.g., stable isotope probing (SIP) approaches (for review see [34]). In this approach a ^{13}C-labeled substrate is added to an environmental isolate or sample, and biomarkers are purified and analysed following the consumption and incorporation of the ^{13}C-labeled substrate into biomass. RNA-SIP and DNA-SIP have been successfully applied for identifying microorganisms responsible for certain biochemical processes, e.g., for the discovery of microbial key players in anaerobic benzene degradation [13, 17, 22, 40]. Protein-based SIP experiments have been recently shown to provide information regarding metabolic activities as well as phylogenetic information [15]. However, this approach is only feasible in pure or enrichment cultures and relies on genome data [48]. Another challenge of SIP approaches is that substrate incorporation has to be sufficient to distinguish between labelled and unlabelled molecules. Groundwater ecosystems are characterised by microbial communities with low cell numbers and low growth rates resulting in low carbon incorporation rates, thus hampering the success of SIP experiments. Furthermore, SIP experiments are relatively time-consuming, and suitable isotopically labelled substrates (e.g., ^{13}C, ^{15}N) are required, which can be expensive. Real-time polymerase chain reaction (qPCR) for quantification of functional genes encoding key enzymes of degradation pathways has also been applied for studying biodegradation potentials at contaminated field sites (e.g., [3]). However, a crucial prerequisite for employing this approach is the knowledge of the degradation pathway of the contaminant and its underlying genetics, which is not the case for all common contaminants.

Another general approach is based on cultivation independent single cell analytics as was used in this study. This may involve various microscopic techniques like fluorescence microscopy, confocal microscopy, and flow cytometry (for review see [33]). The general principle is to analyse bacterial cells one by one according to distinct intrinsic optical characteristics like scatter signals or autofluorescence or to use extrinsic information like fluorescent dyes for labelling of distinct cell features. In this case, various structural and functional characteristics can be visualised and quantified using fluorescent probes. These fluorescent probes can be used for phylogenetic differentiation (FISH, CARD-FISH, RING-FISH targeting rRNA), for the determination of the physiological/functional state of the microorganisms (DNA, storage products, also functional Card-FISH for microscopic application) or various viability states of single cells. However, the combination of fluorescent single cell techniques for phylogenetic differentiation and those that give information on viability and function cannot be applied till now by using flow cytometry alone. Although many methods are available for measuring functions in pure cultures by application of, e.g., fluorescent substrates [33], this cannot be easily applied to environmental samples since specific information cannot be obtained with such approaches to date. Therefore, microscopic approaches like microautoradiography [39], Raman spectroscopy [14] and Nano-SIMS [21] were developed where radioactive or isotopically labelled substrates are combined with in situ hybridisation to identify key players in microbial communities. However, quantification of the key players and following the dynamics of the accompanying organisms remain difficult with these microscopic approaches.

Cell sorting according to fluorescently labelled DNA contents and light scatter characteristics, which indicate multiplication and cell growth, followed by phylogenetic analysis of the separated cells is reliable, robust, and quick. Repeatedly, the approach was shown to generate knowledge of key players in microbial communities [12, 19, 31]. Preconditions are distinct sub-communities and yields of at best 10^6 sorted cells for convenient DNA extraction and PCR. In our case, the limitation was the availability of cell material. Only a few beads were available to extract enough cells for recording flow-cytometric DNA patterns as well as DNA for PCR, cloning, and subsequent T-RFLP analysis. Consequently, nearly no material was left over for the sorting approach. Therefore, only two sub-communities from microcosms exposed in well 52/03 were sorted. The results nevertheless point at specific emerging key organisms. As illustrated in Fig. 8, gate 2 comprised the same *Peptococcaceae* phylotype present also in the whole community, but at a significantly higher proportion. Interestingly, the *Achromobacter* phylotype dominating the whole community was not present in the sorted sub-community, confirming the specificity of the cell-sorting approach. As discussed above, the *Peptococcaceae* might colonise anaerobic micro-niches that occurred within the microcosms after substrates were locally exhausted by aerobic or nitrate-reducing BTEX degradation. The cytometric pattern confirms this succession hypothesis as it shows the emergence of the *Peptococcaceae* from the background of the whole community. These *Peptococcaceae* can be quantified via cytometric gate setting and its abundance followed over longer time ranges. This will facilitate understanding its role in the community. This one or any other organism followed by flow cytometry, cell sorting, and T-RFLP analysis can be seen or defined as a marker organism that gives information on specific micro-environmental conditions at distinct sites and therefore can serve for monitoring and control means. It is the intention of the authors to test the reliability of this approach in more elevated studies in the future.

6 Conclusion

Microbial communities in engineered natural environments can be successfully monitored by studying variations in the composition and structure of the indigenous members with high resolution and quantitatively by using the combined approach of molecular tools, cytometric techniques and appropriate statistical analyses. When a contaminated site is thoroughly studied using this combined approach over a certain time frame, only the quick and inexpensive cytometric pattern analysis together with the Dalmatian plot evaluation may be adequate to monitor microbial community dynamics and predict changes according to biostimulation treatments. Single cell analytics are therefore about to become an indispensable tool when ecosystem stability or instability needs to be investigated on the microbial level.

Acknowledgments This work is integrated in the internal research and development program of the UFZ as well as the SAFIRA project (Remediation research in regionally contaminated aquifers) and the CITE program (Chemicals in the environment). We acknowledge Annett Ohlendorf and Christine Süring for assistance in development of the cell detachment procedure, Werner Kletzander for assistance with the field work, Ute Lohse for technical assistance with the molecular biological analysis and Helga Engewald for technical assistance in the flow cytometric laboratory. We are grateful to Kenneth Wasmund for proofreading and Lorenz Adrian for discussing our manuscript.

References

1. Abdo Z, Schüette UME, Bent SJ, Williams CJ, Forney LJ, Joyce P (2006) Statistical methods for characterizing diversity of microbial communities by analysis of terminal restriction fragment length polymorphisms of 16S rRNA genes. Environ Microbiol 8:929–938

2. Aksoy M (1985) Benzene as a leukemogenic and carcinogenic agent. Am J Ind Med 8:9–20
3. Beller HR, Kane SR, Legler TC, McKelvie JR, Sherwood Lollar B, Pearson F, Balser L, MacKay DM (2008) Comparative assessments of benzene, toluene, and xylene natural attenuation by quantitative polymerase chain reaction analysis of a catabolic gene, signature metabolites, and compound-specific isotope analysis. Environ Sci Technol 42:6065–6072
4. Bombach P, Chatzinotas A, Neu TR, Kästner M, Lueders T, Vogt C (2010a) Enrichment and characterization of a sulfate-reducing toluene-degrading microbial consortium by combining in situ microcosms and stable isotope probing techniques. FEMS Microbiol Ecol 71:237–246
5. Bombach P, Richnow HH, Kästner M, Fischer A (2010b) Current approaches for the assessment of in situ biodegradation. Appl Microbiol Biotechnol 86:839–852
6. Clarke KR (1993) Non-parametric multivariate analyses of changes in community structure. Austral Ecol 18:117–143
7. Da Silva MLB, Alvarez PJJ (2004) Enhanced anaerobic biodegradation of benzene-toluene-ethylbenzene-xylene-ethanol mixtures in bioaugmented aquifer columns. Appl Environ Microbiol 70:4720–4726
8. EEA (2007) Progress in management of contaminated sites. European Environment Agency
9. Foght J (2008) Anaerobic biodegradation of aromatic hydrocarbons: pathways and prospects. J Mol Microbiol Biotechnol 15:93–120
10. Geyer R, Peacock AD, Miltner A, Richnow HH, White DC, Sublette KL, Kästner M (2005) In situ assessment of biodegradation potential using biotraps amended with ^{13}C-labeled benzene or toluene. Environ Sci Technol 39:4983–4989
11. Günther S, Hübschmann T, Rudolf M, Eschenhagen M, Röske I, Harms H, Müller S (2008) Fixation procedures for flow cytometric analysis of environmental bacteria. J Microbiol Methods 75:127–134
12. Günther S, Trutnau M, Kleinsteuber S, Hause G, Bley T, Röske I, Harms H, Müller S (2009) Dynamics of polyphosphate-accumulating bacteria in wastewater treatment plant microbial communities detected via DAPI (4′,6′-diamidino-2-phenylindole) and tetracycline labeling. Appl Environ Microbiol 75:2111–2121
13. Herrmann S, Kleinsteuber S, Chatzinotas A, Kuppardt S, Lueders T, Richnow HH, Vogt C (2010) Functional characterization of an anaerobic benzene-degrading enrichment culture by DNA stable isotope probing. Environ Microbiol 12:401–411
14. Huang WE, Stoecker K, Griffiths R, Newbold L, Daims H, Whiteley AS, Wagner M (2007) Raman-FISH: Combining stable-isotope Raman spectroscopy and fluorescence in situ hybridization for the single cell analysis of identity and function. Environ Microbiol 9:1878–1889
15. Jehmlich N, Schmidt F, Taubert M, Seifert J, von Bergen M, Richnow HH, Vogt C (2009) Comparison of methods for simultaneous identification of bacterial species and determination of metabolic activity by protein-based stable isotope probing (protein-SIP) experiments. Rapid Commun Mass Spectrom 23:1871–1878
16. Jindrova E, Chocova M, Demnerova K, Brenner V (2002) Bacterial aerobic degradation of benzene, toluene, ethylbenzene and xylene. Folia Microbiol 47:83–93
17. Kasai Y, Takahata Y, Manefield M, Watanabe K (2006) RNA-based stable isotope probing and isolation of anaerobic benzene-degrading bacteria from gasoline-contaminated groundwater. Appl Environ Microbiol 72:3586–3592
18. Kästner M, Fischer A, Nijenhuis I, Geyer R, Stelzer N, Bombach P, Tebbe CC, Richnow HH (2006) Assessment of microbial in situ activity in contaminated aquifers. Eng Life Sci 6:234–251
19. Kleinsteuber S, Riis V, Fetzer I, Harms H, Müller S (2006) Population dynamics within a microbial consortium during growth on diesel fuel in saline environments. Appl Environ Microbiol 72:3531–3542
20. Lane DJ (1991) 16S/23S rRNA sequencing. In: Stackebrandt E, Goodfellow M (eds) Nucleic acid techniques in bacterial systematics. Wiley, Chichester, pp 11–175

21. Li T, Wu T-D, Mazéas L, Toffin L, Guerquin-Kern J-L, Leblon G, Bouchez T (2008) Simultaneous analysis of microbial identity and function using NanoSIMS. Environ Microbiol 10:580–588
22. Liou JS-C, DeRito CM, Madsen EL (2008) Field-based and laboratory stable isotope probing surveys of the identities of both aerobic and anaerobic benzene-metabolizing microorganisms in freshwater sediment. Environ Microbiol 10:1964–1977
23. Lonergan DJ, Jenter HL, Coates JD, Phillips EJP, Schmidt TM, Lovley DR (1996) Phylogenetic analysis of dissimilatory Fe(III)-reducing bacteria. J Bacteriol 178:2402–2408
24. Lovley DR, Phillips EJP (1988) Novel mode of microbial energy-metabolism—organic-carbon oxidation coupled to dissimilatory reduction of iron or manganese. Appl Environ Microbiol 54:1472–1480
25. Lovley DR (2001) Bioremediation: anaerobes to the rescue. Science 293:1444–1446
26. Madsen EL (2005) Identifying microorganisms responsible for ecologically significant biogeochemical processes. Nat Rev Microbiol 3:439–446
27. Maher N, Dillon HK, Vermund SH, Unnasch TR (2001) Magnetic bead capture eliminates PCR inhibitors in samples collected from the airborne environment, permitting detection of *Pneumacystis carinii* DNA. Appl Environ Microbiol 67:449–452
28. Mandelbaum RT, Shati MR, Ronen D (1997) In situ microcosms in aquifer bioremediation studies. FEMS Microbiol Rev 20:489–502
29. Mou XZ, Moran MA, Stepanauskas R, Gonzalez JM, Hodson RE (2005) Flow-cytometric cell sorting and subsequent molecular analyses for culture-independent identification of bacterioplankton involved in dimethylsulfoniopropionate transformations. Appl Environ Microbiol 71:1405–1416
30. Mueller S et al. (2006) Differences in fecal microbiota in different European study populations in relation to age, gender, and country: a cross-sectional study. Appl Environ Microbiol 72:1027–1033
31. Müller S, Vogt C, Laube M, Harms H, Kleinsteuber S (2009) Community dynamics within a bacterial consortium during growth on toluene under sulfate-reducing conditions. FEMS Microbiol Ecol 70:586–596
32. Müller S, Harms H, Bley T (2010) Origin and analysis of microbial population heterogeneity in bioprocesses. Curr Opin Biotechnol 21:100–113
33. Müller S, Nebe-von-Caron G (2010) Functional single-cell analyses: Flow cytometry and cell sorting of microbial populations and communities. FEMS Microbiol Rev 34:554–587
34. Neufeld J, Dumont M, Vohra J, Murrell J (2007) Methodological considerations for the use of stable isotope probing in microbial ecology. Microb Ecol 53:435–442
35. Nielsen DR, McLellan PJ, Daugulis AJ (2006) Direct estimation of the oxygen requirements of *Achromobacter xylosoxidans* for aerobic degradation of monoaromatic hydrocarbons (BTEX) in a bioscrubber. Biotechnol Lett 28:1293–1298
36. Oksanen J, Blanchet FG, Kindt R, Legendre P, O'Hara RG, Simpson GL, Solymos P, Stevens MHH, Wagner H (2010) vegan: Community ecology package. R package version 1.17-1. http://cran.R-project.Org/package=vegan
37. Peacock AD, Chang YJ, Istok JD, Krumholz L, Geyer R, Kinsall B, Watson D, Sublette KL, White DC (2004) Utilization of microbial biofilms as monitors of bioremediation. Microb Ecol 47:284–292
38. Porter J, Edwards C, Morgan JAW, Pickup RW (1993) Rapid, automated separation of specific bacteria from lake water and sewage by flow-cytometry and cell sorting. Appl Environ Microbiol 59:3327–3333
39. Rosselló-Mora R, Lee N, Antón J, Wagner M (2003) Substrate uptake in extremely halophilic microbial communities revealed by microautoradiography and fluorescence in situ hybridization. Extremophiles 7:409–413
40. Sakai N, Kurisu F, Yagi O, Nakajima F, Yamamoto K (2009) Identification of putative benzene-degrading bacteria in methanogenic enrichment cultures. J Biosci Bioeng 108:501–507

41. Schirmer M, Dahmke A, Dietrich P, Dietze M, Gödeke S, Richnow HH, Schirmer K, Weiß H, Teutsch G (2006) Natural attenuation research at the contaminated megasite Zeitz. J Hydrol 328:393–407
42. Scow KM, Hicks KA (2005) Natural attenuation and enhanced bioremediation of organic contaminants in groundwater. Curr Opin Biotechnol 16:246–253
43. Spormann AM, Widdel F (2001) Metabolism of alkylbenzenes, alkanes, and other hydrocarbons in anaerobic bacteria. Biodegradation 11:85–105
44. Strevett K, Davidova I, Suflita JM (2002) A comprehensive review of the screening methodology for anaerobic biodegradability of surfactants. Rev Environ Sci Biotechnol 1:143–167
45. Talbot G, Topp, E, Palin MF, Masse DI (2008) Evaluation of molecular methods used for establishing the interactions and functions of microorganisms in anaerobic bioreactors. Water Res. 42:513–537
46. US-EPA (2004) Cleaning up the nation's waste sites: Markets and technology trends. Office of Solid Waste and Emergency Response, Washington, DC
47. van Agteren MH, Keuning S, Janssen DB (1998) Handbook on biodegradation and biological treatment of hazardous organic compounds. Kluwer Academic Publishers, Dordrecht
48. VerBerkmoes NC, Hervey WJ, Shah M, Land M, Hauser L, Larimer FW, Van Berkel GJ, Goeringer DE (2005) Evaluation of "Shotgun" Proteomics for identification of biological threat agents in complex environmental matrixes: experimental simulations. Anal Chem 77:923–932
49. Vieth A, Kästner M, Schirmer M, Weiss H, Gödeke S, Meckenstock RU, Richnow HH (2005) Monitoring in situ biodegradation of benzene and toluene by stable carbon isotope fractionation. Environ Toxicol Chem 24:51–60
50. Vogt C, Lösche A, Kleinsteuber S, Müller S (2005) Population profiles of a stable, commensalistic bacterial culture grown with toluene under sulphate-reducing conditions. Cytometry Part A 66A:91–102
51. Wang YY, Hammes F, Boon N, Chami M, Egli T (2009) Isolation and characterization of low nucleic acid (LNA)-content bacteria. ISME J 3:889–902
52. Wiacek C, Mueller S, Benndorf D (2006) A cytomic approach reveals population heterogeneity of *Cupriavidus necator* in response to harmful phenol concentrations. Proteomics 6:5983–5994
53. Wiedemeier TD, Rifai HS, Newell CJ, Wilson JT (1999) Natural attenuation of fuels and chlorinated solvents in the subsurface. Wiley Inc., London

Adv Biochem Engin/Biotechnol (2011) 124: 183–209
DOI: 10.1007/10_2010_80

Published Online: 11 November 2010

Multivariate Data Analysis Methods for the Interpretation of Microbial Flow Cytometric Data

Hazel M. Davey and Christopher L. Davey

Abstract Flow cytometry is an important technique in cell biology and immunology and has been applied by many groups to the analysis of microorganisms. This has been made possible by developments in hardware that is now sensitive enough to be used routinely for analysis of microbes. However, in contrast to advances in the technology that underpin flow cytometry, there has not been concomitant progress in the software tools required to analyse, display and disseminate the data and manual analysis, of individual samples remains a limiting aspect of the technology. We present two new data sets that illustrate common applications of flow cytometry in microbiology and demonstrate the application of manual data analysis, automated visualisation (including the first description of a new piece of software we are developing to facilitate this), genetic programming, principal components analysis and artificial neural nets to these data. The data analysis methods described here are equally applicable to flow cytometric applications with other cell types.

Keywords Artificial neural nets · Data analysis methods · Flow cytometry · Genetic programming

Contents

H. M. Davey (✉) and C. L. Davey
Institute of Biological, Environmental and Rural Sciences, Aberystwyth University, Penglais, Aberystwyth SY23 3DD, UK
e-mail: hlr@aber.ac.uk

1 Introduction

Flow cytometry offers many advantages for the analysis of microorganisms including the rapid analysis of cells at the single cell level [35] permitting the measurement of heterogeneity [1] and the ability to physically separate cells based on their measured properties. These advantages have been reviewed extensively elsewhere [11, 21, 36]; however, commonly cited drawbacks to the technique include the cost of instrumentation, the requirement for skilled operators and the interpretation of the large volumes of data that are produced. The first of these problems—the expense—is beginning to be addressed through a new generation of instruments that offer adequate numbers of parameters and sufficient sensitivity for microbial analysis in lower cost, sometimes portable machines. The other two problems are linked in that in addition to maintaining, setting up and running the flow cytometer, one of the important skills of the operator is frequently the ability to interpret the results of the experiment. This article seeks to describe the process of manipulation and analysis of flow cytometric data and, via the use of appropriate data sets, to illustrate where computational tools may help alleviate the problem of data handling and interpretation.

1.1 What do Flow Cytometric Data Sets Look Like?

In a typical microbial flow cytometry experiment, a number of samples will be analysed, and ideally the data set will also contain replicate samples and appropriate controls [28, 38]. For example, an experiment may involve samples taken at different stages of a batch culture, or they may be samples of a microorganism that have been exposed to different levels of a stress condition. The results from the analysis of each of the samples will be stored in a separate 'listmode' file, and a set of these files from an experiment (the data set) will typically all be subjected to the same post-acquisition processing to allow similarities and/or differences between the samples to be identified and conclusions to be drawn from the experiment.

Listmode data files conform to a standard that provides the specifications needed to describe the data completely. The first standard (FCS 1) was established in 1984 [37] and is currently [49] at version 3.1. The aim of the standard is to provide a file format that allows files created by one type of acquisition hardware and software to be analysed by any other type and has the added advantage that third party or end-user created software packages can read the files. The files consist of a header and a data area. The header has information about the identity of the sample, the flow cytometer settings used to run the sample, etc., while the data area records measurements of each particle (e.g. microbial cell or spore) detected by the flow cytometer.

When a particle passes through the detection point of the flow cytometer, it intersects a beam of light from a laser and/or arc lamp. When this happens, light is scattered out of the incident beam. Light scattering provides information on the size and structure of the cells, but the relationship is far from straightforward [36] particularly when cells of different types are considered. Additionally, fluorescence may occur and be measured at a number of discrete wavelengths. In some cases, e.g. the analysis of phytoplankton, this may be autofluorescence from the photosynthetic pigments present in the cells, more commonly though the fluorescence will result from fluorescent stains, fluorogenic substrates or fluorescently labelled antibodies that are added to the sample prior to analysis. The data area of the listmode file will thus consist of a number of values for each recorded particle that correspond to the light scattering or fluorescence intensity measured by the appropriate detectors. The intensities are measured on an arbitrary scale of channel numbers that may be distributed on a linear or log scale depending on the way that the flow cytometer was set up to acquire the samples. Flow cytometers typically analyse at very rapid rates—of the order of several hundred to more than a thousand particles per second. Thus, it can be seen that each file in the flow cytometric data set will contain multiple measurements made on tens of thousands or hundreds of thousands of individual particles. This represents a challenge for handling, storage and analysis of the data produced.

2 Methods: Generation of Flow Cytometric Data Sets

In order to illustrate the methods that can be used to analyse flow cytometric data, two data sets were acquired according to the following methods and rationale.

2.1 Data Set 1: Detection of a Specific Microbial Threat

In recent years there has been increased concern regarding the threat of bioterrorism, and reliable techniques are required to detect and identify bioagents in order that their impact on public health can be minimised [29]. With its short incubation time, high mortality rate and environmental resistance, *Bacillus anthracis*, the causative organism of anthrax, has long been considered the most credible biowarfare threat [8]. The anthrax attacks in the US in 2001 confirmed the status of this pathogen in the modern context of bioterrorism [5, 22]. Rapid and accurate determination of the presence, concentration and identity of microbial cells in environmental samples is made more difficult by the complex, particulate background against which the target organism must be identified. Flow cytometry has the advantage that particles are measured individually, but such monitoring would be most effective if the results were available in (near) real-time, during or immediately after sample analysis, as this would permit the use of sensors to trigger defensive strategies and medical interventions. A rapid first-stage screen with low-cost, non-specific reagents would permit reanalysis of the sample to confirm the results or the performance of specific tests—e.g. via the use of fluorescently labelled antibodies. Additionally, in suitably equipped instruments, detection and identification during analysis permits physical separation (sorting) of the target organism to enrich it into a small volume, thereby increasing the efficiency and economic viability of specific tests. For this to take place, efficient and automated analysis is required.

A data set was therefore generated following methods previously outlined [10] with *Bacillus globigii* (*B. subtilis* var niger, a well-established simulant for *B. anthracis* [50]) as the target to be identified against a background of other microbial particles. *B. globigii* spores were obtained from the Chemical and Biological Defence Establishment, Porton Down, Salisbury, UK, as a dry preparation. Prior to analysis the spores were suspended in sheath fluid (see below) to give a concentration of approximately 1×10^6 spores ml^{-1}. *Escherichia coli* (Lab Strain C500), *Micrococcus luteus* (NCIMB 13267) and *Saccharomyces cerevisiae* were grown as previously described [10]. In order to deactivate microorganisms and to facilitate storage, fixed samples were prepared by passing cell or spore suspensions through a fine needle into ice-cold absolute ethanol to give a final ethanol concentration of 70%. Prior to analysis, the fixative was removed by centrifugation, and the cells were washed and resuspended in sheath fluid at an appropriate concentration. Mixed samples were prepared by combining the organisms to give known numbers of each cell type. Unstained controls were analysed without further treatment. Stained samples were prepared using a cocktail of three fluorescent stains—Tinopal CBS-X (Ciba Dyes and Chemicals Ltd., Macclesfield, UK) at a final concentration of 40 $\mu g\ ml^{-1}$, propidium iodide (Sigma, UK) at 50 $\mu g\ ml^{-1}$ and fluorescein isothiocyanate (Sigma, UK) at 25 $\mu g\ ml^{-1}$. Propidium iodide targets nucleic acids and fluorescein isothiocyanate binds to proteins. The exact target of Tinopal CBS-X has not been determined. Multiparametric flow cytometric analyses were performed using a Coulter Epics Elite flow cytometer (Beckman Coulter UK Limited). Samples were illuminated using a helium-cadmium laser (325 nm) and an argon ion laser (488 nm), and the gated-amp assembly was used to recombine the signals from the separate lasers. The flow cytometer was set up as described in the manufacturer's manual and 10 μm fluorescent beads were used for alignment. The sheath fluid was prepared

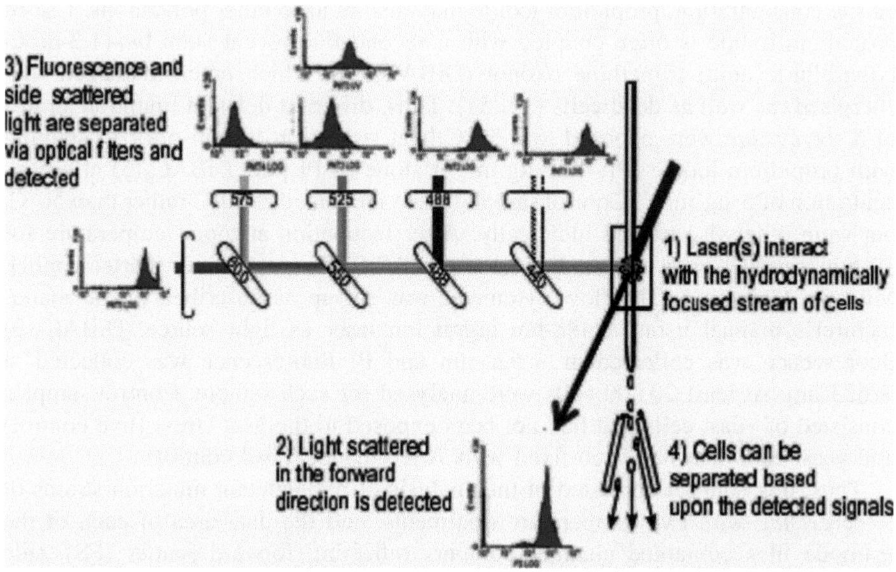

Fig. 1 Flow cytometric analysis involves a system of fluidics for sample delivery and optics for illumination and measurement of the sample. Scattered and fluorescent light is detected by photomultiplier tubes (PMTs) for each particle in the sample. The graphs show analysis of a sample of *S. cerevisiae* from data set 1. Two signals are collected by PMT3 (~525 nm)—one from the UV laser and one from the argon ion laser

using Millipore Milli-Q water filtered to 0.22 μm and contained 150 mM KCl and 10 mM HEPES. The pH was adjusted to 6.8 with KOH and the sheath fluid was then filtered to 0.1 μm using a Whatman WCN filter. The forward scatter (FS) signal from the argon ion laser was used to discriminate between the signal from the microorganisms and the background noise. The data files typically consisted of ~10,000 particles per file with seven parameters (Fig. 1) being measured from each particle.

2.2 Data Set 2: Determination of Viability

The aim of the second experiment was to prepare a data set that would illustrate the use of flow cytometry to measure viability in *Saccharomyces cerevisiae*. This yeast is used in biotechnology and is also an important model organism [16]. As such, methods for routine monitoring of its physiology are important. A widely used stain for this purpose is propidium iodide, which acts as an effective viability indicator in that the propidium ion is excluded by the intact membrane of viable yeast cells but freely enters cells with damaged membranes whereupon it binds to nucleic acids resulting in an enhancement of the stain's fluorescence. Depending on the concentration, propidium iodide may also bind to other polyanions. Use of propidium iodide is often coupled with a second fluorescent stain bis-(1,3-dibutylbarbituric acid) trimethine oxonol ($DiBAC_4(3)$), which enters damaged (de-energised) as well as dead cells [13, 54]. Thus, different deletant mutation strains of *S. cerevisiae* were exposed to a 50°C heat stress for 30 min prior to staining with propidium iodide (PI; 4.66 μg ml^{-1}) alone or PI plus $DiBAC_4(3)$ at a concentration of 5 μg ml^{-1}. Control samples were incubated at 30°C rather than 50°C, but were otherwise treated identically. After incubation at room temperature for 10 min, samples were analysed on a Partec PAS III flow cytometer (Partec GmbH, Münster, Germany). The flow cytometer was set up as described in the manufacturer's manual using a 488-nm argon ion laser as light source. $DiBAC_4(3)$ fluorescence was collected at ~520 nm and PI fluorescence was collected at ~633 nm. At least 20,000 cells were analysed for each sample. Control samples consisted of yeast cells that had not been exposed to the heat stress (live control) and yeast cells that had been fixed with 70% ethanol (dead control).

Thus, this data set consisted of the analysis of five deletant mutation strains of *S. cerevisiae* with two temperature treatments, and the data area of each of the listmode files contained channel numbers reflecting forward scatter (FS), side scatter (SS), green fluorescence (520 nm) and red fluorescence (633 nm). Additionally, as a standard template was used, the flow cytometer automatically measured and stored orange fluorescence (~580 nm) and deep red fluorescence (~665 nm), and, although not expected to encode additional information, these parameters were used in multiparametric analyses (see later).

3 Data Visualisation

In order to be able to interpret the data, the first step is usually to produce a graph (or graphs) of the data that have been acquired. Flow cytometers are supplied with software that is used both to control the instrument and to display the data during acquisition. Screen shots of such software can often be seen on the websites of the manufacturing companies. Typically the data display is a panel of single and dual parameter plots that show the data in a way that can be customised by the operator according to the needs of the experiment and to his or her personal preferences (Fig. 2). As particles pass through the flow cytometer, the magnitude of pulses representing the extent of light scattered or the intensity of fluorescence at a given wavelength is sorted electronically into "bins" or "channels," permitting the display of histograms of the number of cells possessing a certain quantitative property versus the channel number. Three parameters can be plotted on a single 3D-scatter plot, but these graphs can be difficult to interpret, particularly when represented as a 2D print. Of course, even with 3D plots, information is lost, but it is generally impractical to visualise more than three dimensions simultaneously.

As shown in the examples given above, even with microbial flow cytometry it is possible to collect data on seven different parameters for each cell, albeit that two

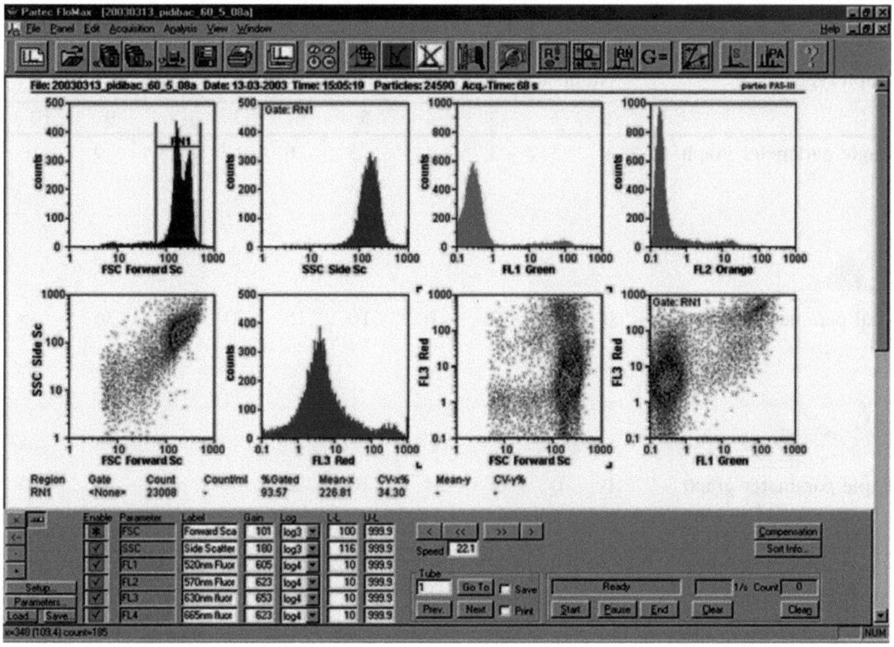

Fig. 2 Acquisition of flow cytometric data is typically carried out in software provided by the manufacturer, in this example Partec (Partec GmbH, Münster, Germany). The data display can be customised by the operator to display the required combinations of parameters. This process allows subpopulations to be identified, such as the population of dead cells in the bottom-right graph that have taken up both propidium iodide and DiBAC$_4$(3)

or three dyes (4–5 parameters) are more common. Higher numbers of parameters are more frequently associated with the multicolour analyses used in immunophenotyping [2, 6, 44] and in analysis of fluorescence of aquatic samples [20], but any analysis that involves more than three parameters presents a problem when it comes to displaying the data in a useful way that will allow relationships to be observed.

As described above it is usually desirable to display the data in two or, at most, three dimensions. However, using this approach means that as the number of measured parameters (n) increases, one must examine more graphs (G) to explore the data. As shown in Table 1, whilst the number of single parameter graphs that need to be examined increases by one for each additional parameter acquired, the number of 2D and 3D plots rapidly becomes unmanageable. Instruments capable of up to 18-colour analysis have found particular utility in immunophenotyping studies where a range of markers exist for distinguishing between cell types [3, 17, 34]. An instrument capable of collecting up to 62 parameters has recently been launched (http://www.isac-net.org/index.php?option=com_content&task=view&id=795&Itemid=1), and procedures have been proposed that allow multiple analyses to be combined, creating data files of potentially infinite dimensions

Table 1 The number of graphs of different types that need to be examined depends upon the dimensionality of the data and the plotting method used

Graph type	Number of parameters measured									
	1	2	3	4	5	6	7	8	9	10
Single parameter graph $G=n$ 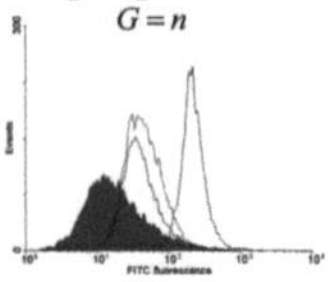	1	2	3	4	5	6	7	8	9	10
Dual parameter graph $G=\frac{n(n-1)}{2}$ 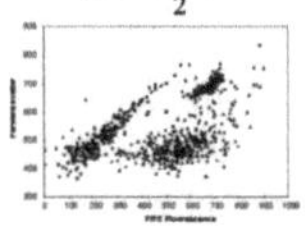	0	1	3	6	10	15	21	28	36	45
Triple parameter graph $G=\frac{n(n-1)(n-2)}{6}$ 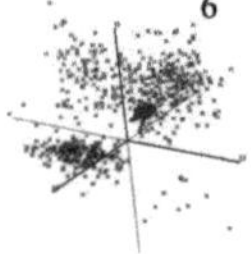	0	0	1	4	10	20	35	56	84	120

[40]. The unprecedented dimensionality of these data sets means that a full exploration of the data space by generation of graphs is clearly impractical.

For straightforward analyses of a small number of parameters, the acquisition software may be sufficient, but for more complex analyses or for production of publication-quality graphs, additional software is often used. Off-line analysis away from the computer on which the data were collected is advantageous in that it frees up the flow cytometer for acquisition of data from other samples. The flow cytometric community has developed a range of software, much of which has been made available free of charge via links available at http://www.cyto.purdue.edu/flowcyt/software/Catalog.htm. Whether free or commercial, a frequently encountered problem is the lack of ongoing support or development, meaning that older software cannot read the latest data files or use the longer file naming conventions on newer operating systems.

3.1 Manual Data Processing

Manual analysis of flow cytometric data involves using the software tools described earlier to produce the necessary graphs. Usually the operator has preconceived ideas based on the experimental design regarding which combinations of parameters should be plotted together. For example, in the case of data set 1, all

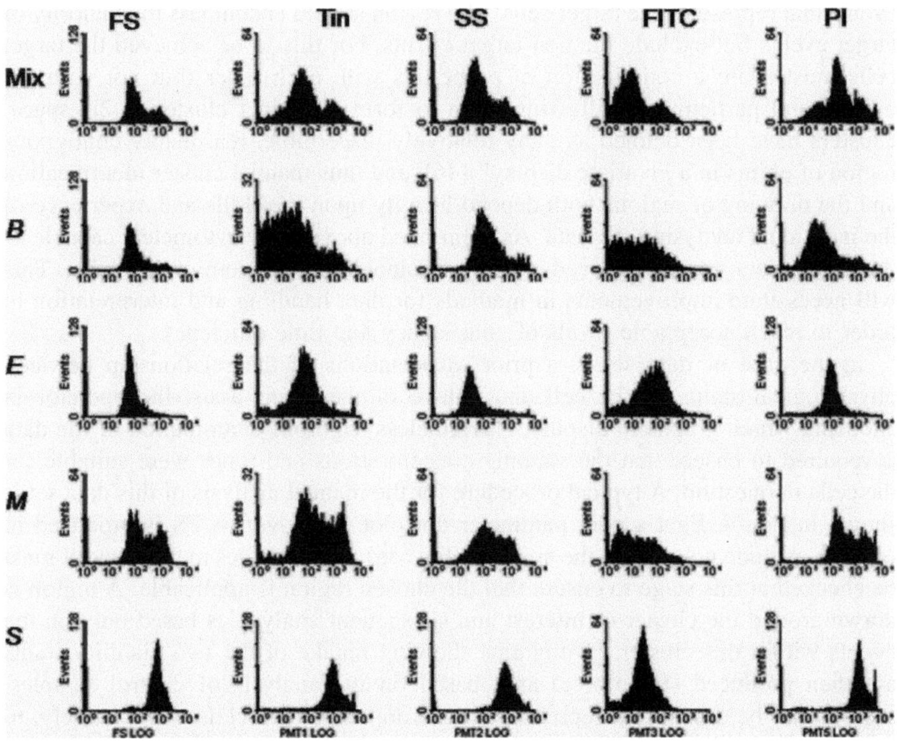

Fig. 3 Single parameter histograms for the four microorganisms (*B*, *B. globigii*; *E*, *E. coli*; *M*, *M. luteus* and *S*, *S. cerevisiae*) plus a mixture (labelled 'mix'). Five of the measured parameters are shown—*FS* forward scatter, *Tin* Tinopal CBS-X at <440 nm, *SS* side scatter, *FITC* fluorescein isothiocyanate at ~525 nm and *PI* propidium iodide at >600 nm. The remaining signals (Tinopal fluorescence at ~525 nm and fluorescence at 575 nm) were highly correlated with other signals and are omitted for clarity. None of the measured parameters permitted effective discrimination of *B. globigii*

single parameter plots would be examined to identify if any of the stains offers effective discrimination between the target organism and the other organisms. Such an analysis of five of the parameters from data set 1 is shown in Fig. 3. In the event that a single parameter offers sufficient discrimination, a region can be drawn on the relevant histogram, and the number of positive events (events falling within the region) can be determined. Frequently however, a second parameter is required, and ideally all combinations of parameters should be examined for each file. With five parameters per sample, this would represent ten 2D dot plots (Table 1) per file—with four organisms plus the mixture, this would be 50 graphs (and this number would scale with the number of replicates). So-called 'stare and compare' analysis is not particularly efficient, but once a suitable pair of parameters has been identified, a region or gate is drawn around a group (cluster) of events that represents the target cells. The region should encompass the majority of target events but exclude the non-target events. For this to be achieved the target cells must share a combination of properties with each other (but not with the background particulates), allowing them to form a distinct cluster in 2D space. Clusters have been defined as "any relatively discernible, reasonably contiguous region of points in a bivariate display" [46], and thus manual cluster identification and the drawing of regions both depend heavily upon the skills and experience of the individual analysing the data. As mentioned above, flow cytometers capable of producing data sets of unprecedented dimensionality are becoming available. This will necessitate improvements in methods for data handling and interpretation in order to reach acceptable levels of consistency and time efficiency.

In the case of data set 2, a priori expectations of the relationship between physiological status of the cell and uptake of the stains assist the operator in choosing which graphs to display. Nevertheless, rigorous examination of the data is required to ensure that the staining concentrations and times were suitable for the cells in question. A typical procedure for the manual analysis of this data set is shown in Fig. 4. First a dual parameter dot plot of SS versus FS is produced in order to exclude noise from the analysis (Fig. 4a). All data files in the data set must be checked at this stage to ensure that the chosen region is applicable. A region is drawn around the cluster of interest and subsequent analysis is based only on the events within this cluster. Histograms showing uptake of the two viability stains are then produced (Fig. 4b, c) and, based on the analysis of control samples, regions can be drawn that represent stained and unstained cells. Alternatively, to exploit the multiparametric nature of flow cytometric data, a second dual parameter graph can be produced of PI vs $DiBAC_4(3)$ fluorescence (Fig. 4d). To aid in interpretation of the data from this graph, quadrants were drawn as shown. The quadrants are drawn on the basis of analysis of control samples such that live, undamaged cells (non-fluorescent) are recorded in the lower left quadrant, live but damaged cells ($DiBAC_4(3)$ positive, PI negative) are in the lower right quadrant, and dead cells [$DiBAC_4(3)$ positive, PI positive] are in the upper right quadrant. This process is then repeated for all strains in the data set (Fig. 4e). The processes of drawing the regions and the quadrants are both manual and based on a 'by eye' optimisation of their positioning and as such the experience of the investigator is extremely important. As a consequence of this, manual processing of flow cytometric data is time-consuming and open to the possibilities of inconsistent analysis and/or inadvertent bias. Nevertheless, this approach remains the route most frequently taken in the analysis of flow cytometric data.

3.2 Development of Automated Visualisation Software: FlowDec

As a consequence of the benefits and drawbacks of manual data analysis explained earlier, we have been developing a programme called "FlowDec" to make the process of manually analysing flow cytometric data sets as straightforward and

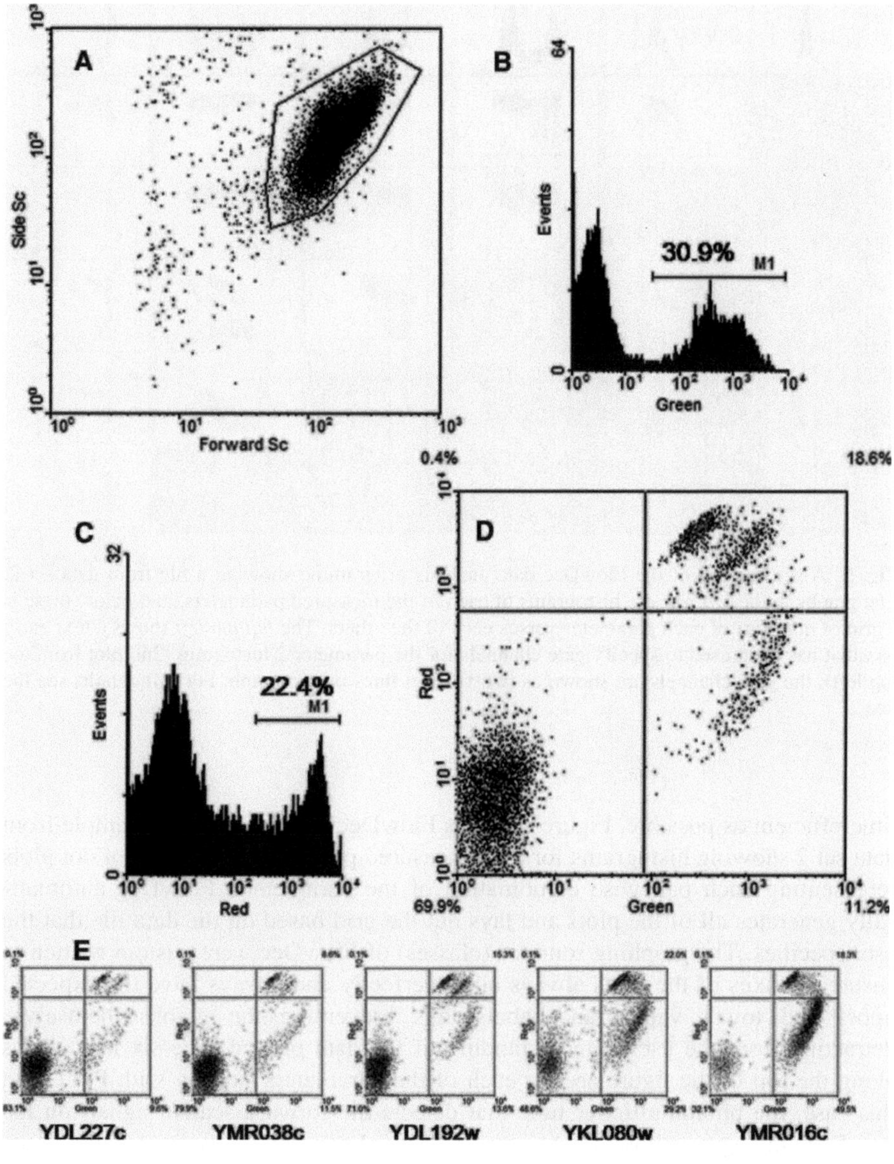

Fig. 4 Sequential steps in the manual analysis of flow cytometric data for determination of the physiological state of *S. cerevisiae*. **a** A dual parameter dot plot of SS versus FS. **b** Histogram showing uptake of DiBAC$_4$(3). On the basis of analysis of control samples, those cells in M1 are considered to have deenergised membranes. **c** Histogram showing uptake of PI; on the basis of analysis of control samples those cells in M1 are assumed to be dead. **d** A dual parameter plot of PI vs DiBAC$_4$(3) fluorescence. Quadrants have been drawn to aid interpretation of the data (see the text for further information). **e** Comparison of yeast strains reveals a lower resistance to heat stress in heterozygous deletant strains of YKL080w and YMR016c when compared to the other strains

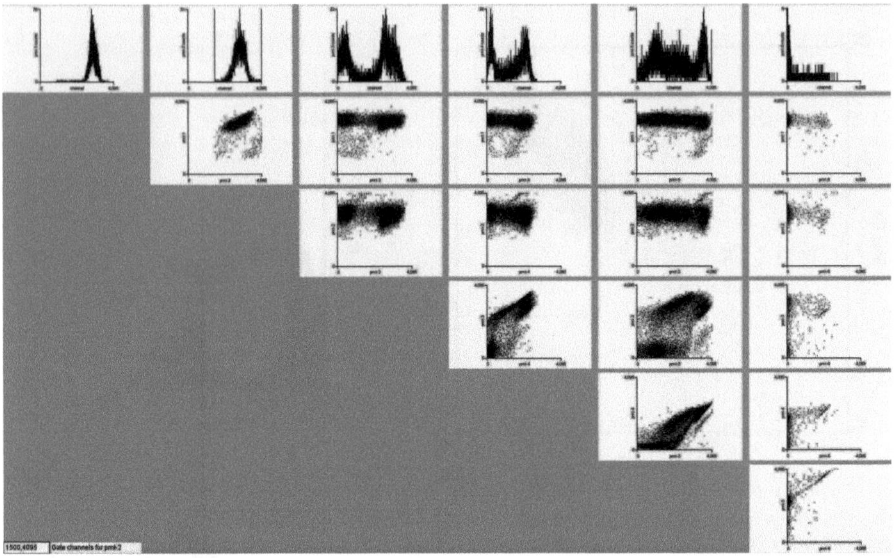

Fig. 5 A screen shot of the FlowDec data analysis programme showing a file from data set 2. The graphs in the *top row* are histograms of each of the measured parameters, and below these is a grid of *dot plots* of each parameter versus each of the others. The *bottom left* shows a text entry box that has been used to specify gate channels for the parameter 2 histogram (2nd plot from the top left); the gate channels are shown as two vertical lines on the graph. For full details, see the text

time-efficient as possible. Figure 5 shows FlowDec's display for one sample from data set 2 showing histograms for each measured parameter and a grid of dot plots representing each pair-wise combination of the parameters. FlowDec automatically generates all of the plots and lays out the grid based on the data file that the user specifies. The graphing routines (classes) of FlowDec were custom written to ensure the axes of the plots always align perfectly and always have the expected upper and lower values and labels thus preventing the graphs themselves detracting from the user's understanding of the data plotted. The six histograms along the top of the figure are for each of the parameters starting with PMT1 (in this case, the photomultiplier tube that detects the forward scatter signal) on the left. Each histogram shows counts versus channel number and is auto-scaled to the largest peak. However, as noise can be a problem in flow cytometric analyses of microorganisms, a facility to ignore certain channels in calculation of the largest peak has been added—in the example shown, data in the first four channels are excluded. Below the histograms is a grid of dot plots. The top row of dot plots has PMT1 channel numbers on the ordinate axes of the graphs, the row below has PMT2 channel numbers on the ordinates and so on. The abscissa for each column of dot plots is for the same parameter shown on the histogram at the top of the column. This enables users to look directly down from the peaks on a histogram and see how the data look when a second dimension is added. Only half of the total grid of dot plots is plotted (i.e. where PMT1 vs. PMT2 is plotted, PMT2 vs. PMT1 is not needed as it conveys no additional information). This is partly because of the time needed to generate large numbers of data-intensive graphs and partly as it reduces the effort needed by the user to comprehend trends in the data. The PMT2 histogram (second graph from the top left) has two vertical bars plotted. These represent the two channel numbers used to gate the data and were specified in the text entry box shown at the bottom left of the figure. When gate channels are entered, FlowDec replots all of the graphs but excludes events that do not fall between the two gate channels. By putting the gate channels on either side of the peaks on the histograms, the user can plot only the events that correspond to each peak.

Thus, FlowDec enables quick and efficient visual exploration of the multiple data files produced by a typical experiment. FlowDec is an ongoing development project that currently has two further aims. The first is to add, on an incremental basis, new analysis features to the programme as the need for them is identified and then to evaluate their usefulness. This ultimately may include genetic algorithms and artificial neural networks (see later) that will, for the first time, permit users to apply these tools to flow cytometric data without manually converting between data formats and using multiple software packages. Secondly, we will use FlowDec as a platform to develop novel visualisations of flow cytometric data that are directly aimed at answering the questions that users are actually asking when they examine the 2D graphs and histograms. Such alternative visualisations of data are now being used to make other large and complex data sets more comprehensible [18, 26, 48], but with flow cytometric data, we envisage their use as an adjunct to, rather than a replacement for, normal manual analysis methods.

This broad range of aims for FlowDec along with the focus on rolling development, generating bespoke visual representations and facilitating interaction, means that we have had to take great care with the computer language chosen to implement the programme. We are using the language "Processing", which is a variation of Sun Java [47]. This was originally developed as a scripting language for artists and animators [43], but has now come to be widely used in the interactive exploration and representation of large scientific data sets [18, 48]. The strengths of Processing for this purpose are that ideas can be explored using scripting and then finalised by being converted to classes and implemented in object-orientated form if required, and that the structure of a Processing programme makes the development of interactivity [47, 52] much easier than would be the case with native Java. In addition, Processing has a rich set of facilities and libraries that can be incorporated into custom-built programmes. For example, these facilitate easier development of applications that involve data parsing, 2D and 3D rendering, animation and sound, and there is now a library for genetic algorithms. As the language is a variation of Java, that language's facilities and libraries are also available to the programmer (e.g. for artificial neural networks, data plotting, graphs and networks, etc.). Like Java, Processing is platform independent, and so it is possible to create versions of the same programme for the web (as an applet) or to run as a normal programme on a computer under different operating systems. These strengths do, however, come with some compromise to the speed of execution of the software. Processing is a free download from http://www.processing.org/, and additionally Java must also be installed—this can be downloaded for free from http://java.com/en/download/index.jsp. The latest version of FlowDec (v1.00) is available from http://qbab.aber.ac.uk/flow/software.html.

3.3 Mathematical Interpretation of Flow Cytometric Data

As can be seen above, whichever approach (manual or automated) is taken to generate graphs of flow cytometric data, interpretation of single parameter histograms is more straightforward than the analysis of dot plots. Effectively what is done is to set a threshold, for example, in the case of data set 2, red fluorescence above a certain channel number indicates that the cell is dead (Fig. 4c). Thus, rather than plotting each data file, a mathematical interpretation of the data becomes possible, which is advantageous for automated detection systems [29]. However, it is the multiparametric nature of flow cytometric data that is a particular advantage of the technique, and the mathematical approach has in a number of cases been extended to incorporate two parameters via the use of ratios. The ratio of two "standard" acquired parameters has been used to estimate membrane potential in bacteria [39] and, more recently, to detect calcium signalling [7]. The ratio approach is also applicable to the case of the problem defined in data set 1 [9] where the ratio of propidium iodide (PI) fluorescence to forward scatter (FS) was found to be discriminatory. Figure 6a shows this ratio calculated from an analysis of a mixed sample containing all four organisms in approximately equal amounts and reveals an encouraging split into two populations. Using prior experience that the majority of fluorescent stains tested stain spores to a lesser extent than is observed with vegetative organisms [10, 12], it was hypothesised that the majority of the *B. globigii* spores were present in the subpopulation with the lowest PI:FS ratio. Analysis of each of the microorganisms in isolation confirmed that the smaller subpopulation identified does indeed largely represent the *B. globigii* spores, whilst the larger population is formed from a combination of the three other organisms (Fig. 6b). This finding allows a simple threshold to be set based on the ratio of the two parameters to provide a first stage identification of *B. globigii*. If a threshold of 1 is set, then 90.5% of events are correctly identified (Fig. 6c) with 9.5% false negatives and 9.5% false positives. Often there is a greater penalty for one type of error, and raising the threshold to 1.049 improves the detection of *B. globigii* (events with a PI:FS ratio of less than 1.049) to 98%, but increases the false-positive rate to 14.5%. One advantage of the mathematical approach over 'stare and compare' analysis is that a sequence of criteria can be applied to each event. Thus, the accuracy of detection of *B. globigii* can be further improved by the use of additional parameters, for example, if the measured data are subjected to the additional criterion that the fluorescein isothiocyanate:side scatter (FITC:SS) ratio for *B. globigii* should be less than 1.06, then there is one additional false negative but nine fewer false positives (not shown). In this way an

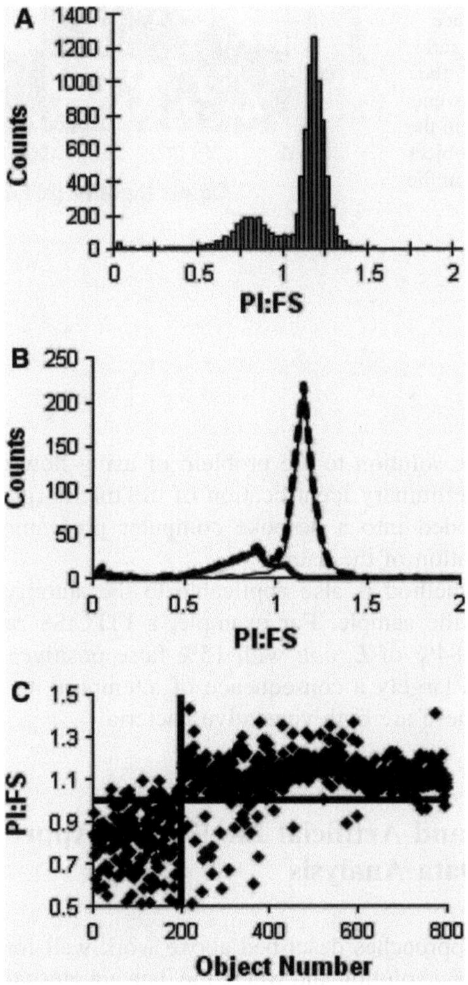

Fig. 6 Ratiometric flow cytometry. **a** Flow cytometric analysis of a mixture of *B. globigii* spores, *E. coli*, *M. luteus* and *S. cerevisiae*. Using the derived ratio of propidium iodide (PI): forward scatter (FS), the population is split into two subpopulations. **b** Analysis of each organism separately confirmed the hypothesis that the subpopulation with a low PI:FS ratio was *B. globigii* (*thick, solid line*, *B. globigii*, *thin line*, summed data from the other three organisms, *dotted line*, mixture of all four organisms). By plotting the PI fluorescence for each cell/spore as a ratio to its forward scatter signal, the majority of the *B. globigii* events can be distinguished from the other microorganisms. **c** A threshold of 1 for the PI:FS ratio allows correct classification of 90.5% of the events as being *B. globigii* (ratio < 1) or not *B. globigii* (ratio > 1). Object numbers 1–200 are *B. globigii*, 201–400 are *E. coli*, 401–600 are *M. luteus* and 601–800 are *S. cerevisiae*

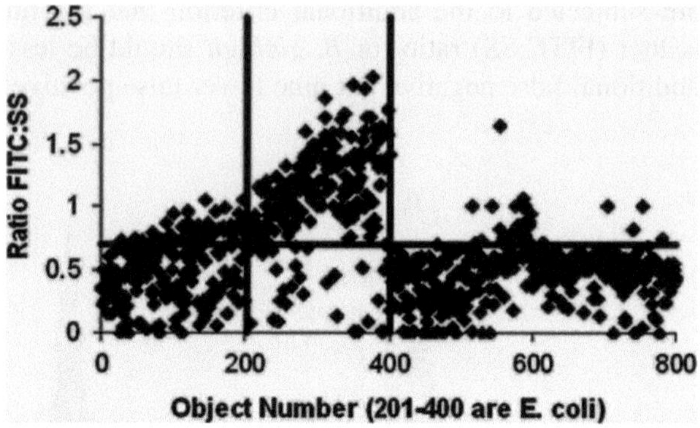

Fig. 7 Ratiometric flow cytometry reveals that a FITC:SS ratio greater than 0.7 correctly discriminates the majority of *E. coli* from the other organisms. By plotting FITC fluorescence for each cell/spore as a ratio to its side scatter signal, the majority of the *E. coli* events can be distinguished from the other microorganisms (object numbers are as defined in the legend to Fig. 6)

effective algorithmic solution to the problem of using flow cytometric measurements to make a preliminary identification of the microorganisms present in the sample could be coded into a bespoke computer programme dedicated to the automated interpretation of the data.

The ratiometric method is also applicable to the automated detection of the other organisms in the sample. For example, a FITC:SS ratio greater than 0.7 correctly described 84% of *E. coli* with 15% false positives (Fig. 7). The lower level of accuracy is largely a consequence of attempting to discriminate *E. coli* from *M. luteus* as these are both vegetative bacteria.

4 Multivariate and Artificial Intelligence Approaches to Flow Cytometric Data Analysis

The mathematical approaches described above work well for data analysis problems that have an exact solution and where the data are close to noise-free. In order for this approach to be effective, software programmers must be able to define all of the steps in the algorithm that need to be carried out to go from the data to the desired output (the correct answer). The algorithm-based software programme is then written to take the inputs and combine them in a specific way through a set of sequential calculations to reach the conclusions. Quite apart from the fact that the correct algorithm may not be apparent, in biology generally and flow cytometry in particular, data are seldom noise-free. Such a set of circumstances requires a different approach to identify an appropriate algorithm and to achieve a robust result.

4.1 Genetic Algorithms and Genetic Programming for Algorithm Development

A genetic algorithm (GA) is an automated computer method to find a solution to a problem. GAs use techniques based in evolutionary biology such as cross over, mutation and selection to evolve a random starting population of "chromosomes" into more accurate and useful solutions. The "fitness" of each individual in the population is evaluated in terms of its ability to match a flow cytometric data pattern to the correct identity of the particle giving rise to the data. The most accurate algorithms are selected and modified by recombination and/or mutation to form the next generation. This process is repeated until errors are minimised and the "best" algorithmic solution is found. Genetic programming (GP) is an extension of GA in which a GA is used to "discover" an optimal set of rules (in the form of a computer programme) by rearranging a set of available parts [19].

To illustrate this approach, a GP was used to classify yeast cells as "live" or "dead" using the free software package A.I. Solver Studio (http://www.perseptio.com). As this is a generic piece of software rather than one designed for use with flow cytometric data, its application to the data is somewhat convoluted. The first step is to convert the data from the flow cytometry standard listmode format into an ASCII text file. The freeware, command line programme lldata.exe (http://www.cyto.purdue.edu/flowcyt/software/Catalog.htm#purdue), was used for this purpose, and this generates a formatted file where each row in the file represents one analysed particle and each column represents a measured parameter with the parameters being separated by multiple spaces. The A.I. Solver software requires a single delimiter between parameters, a label at the beginning of each row that identifies the particle as "live" or "dead" and a header row describing the contents of each column. Thus, to create data files in the correct format, samples of live (overnight cultures) and dead (ethanol fixed) yeast were stained with PI and analysed on the Partec PAS III. The resulting files were converted using lldata.exe, loaded into Microsoft Excel and 100-example patterns of live yeast and 100-example patterns of dead yeast were selected at random and put into a new file. A header row was added, and an additional spreadsheet cell was inserted at the start of each row to identify each of the patterns. These data were partitioned, again randomly, into a 'training' file and a 'test' file. The training file contained data patterns representing 84 dead and 86 live yeast, and the test file contained data corresponding to 16 dead and 14 live yeast. Files were saved as comma-separated variable files. These files were then loaded into AI solver studio, and the software was set up to assume that all errors were of equal consequence, that the solution was of 'medium complexity' and to allow 25% of the training data to be used for over-fitting prevention. Over-fitting prevention is used to eliminate a problem commonly encountered in algorithms that are trained to associate patterns with outputs whereby they achieve this by learning the associations exactly but in the process, lose the ability to generalise. Training progresses and stops automatically once an acceptable error level is achieved—in this case after 1,195 iterations

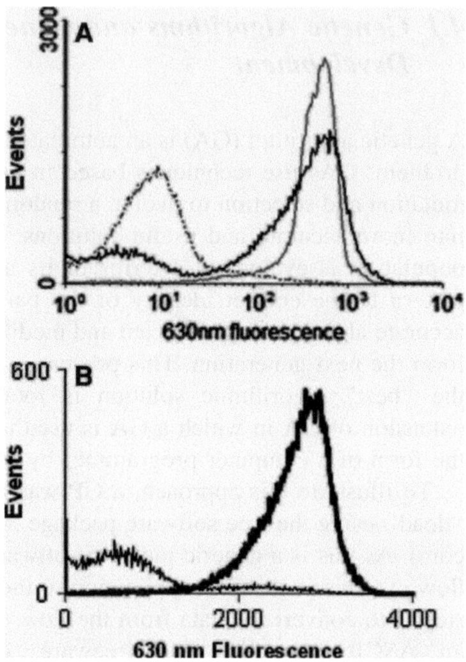

Fig. 8 A genetic programme was developed as described in the text to discriminate between live and dead yeast based on the analysis of controls. **a** Manual data analysis of live (*dotted line*), dead (*thin line*) and heat-stressed (*thick line*) yeast. **b** When data representing 80,446 heat-stressed cells were presented to the GP for classification, red fluorescence was clearly an important factor in the GP rule that distinguished live (*thin line*) from dead (*thick line*). This is in agreement with expectation, but it is not dependent upon the experience of the operator and does not limit the solution to one or two of the measured parameters

(presentations of the training set). At this point all of the training data were correctly classified, and in the test data 88% dead and 92% live yeast examples were correctly identified. An unclassified data set from the analysis of a mixed viability sample was then loaded into the software. This contained 80,446 events that were presented to the trained GP for classification as "live" or "dead". Here of course the physiological status of each particle is unknown (the file represents a mixture of live and dead cells), and the GP needs to make a prediction of identity according to the rules that it has developed. Conventional analysis would be as described earlier (Fig. 4) through the analysis of control samples and the drawing of gates, but with the GP the differentiation is automated and potentially uses all measured parameters. Thus, it is not dependent upon the experience of the operator, does not limit the solution to one or two of the measured parameters, and is not open to bias. The GP classification and manual classification of the data are shown in Fig. 8.

Unfortunately, whilst freeware packages such as AI Solver produce the trained GP and allow its interrogation as has been used above, they do not provide the GP solution to the problem in an equation form. Also, as demonstrated by our example, the data handling involves multiple steps and conversions increasing the chances of errors being introduced, compromising data traceability and slowing down the process considerably. Consequently, there have been very few applications of this technology to flow cytometric data; however, Day et al. [14] used a commercial GP package to differentiate *Phytophthora infestans* sporangia from other particulates (conidia, pollen, etc.), allowing 95% of the sporangia to be correctly identified with a false-positive rate of <1%. The best rule that was evolved involved all measured parameters and contained nonlinear operators, and was thus a very different solution to that which could have been obtained by conventional flow cytometric analysis methods that rely on clusters being linearly separable, usually in 2D space. In theory, as with the more straightforward mathematical algorithms described above, this rule could be coded into a bespoke piece of software to interpret the data.

4.2 Dimension Reduction

An alternative approach to dealing with multiparametric data sets is to reduce the dimensionality to a more easily interpretable number of dimensions [15, 27, 51]. For example, the aim of Principal Components Analysis (PCA) [23, 24] is to rotate the data points into a new co-ordinate system, such that the majority of the variance in the data set is accounted for by a subset of these rotated axes. Hence, by plotting the points in this new coordinate system (rather than the more familiar scatter and fluorescence parameters), the significant differences within the multiparametric data set can be more easily visualised in two or three dimensions.

PCA is applicable to the problem defined in data set 1 (the detection and the identification of a particular microorganism) where the aim is to distinguish microbes from each other. To this end, a subset of the data representing all four organisms was analysed using PCA, and two-dimensional plots were produced of the 1st and 2nd, 1st and 3rd and 2nd and 3rd most discriminatory factors. The plot that gave the best separation of the *B. globigii* data from the other organisms was chosen, and a polygon was drawn around the *B. globigii* events (Fig. 9). The first and third principal components were found to give the best partitioning of the data with 98.5% of *B. globigii* being identified with 3.2% false positives. As with manual analysis of raw flow cytometric data, the positioning, shape and size of the polygon are judged by eye to achieve optimal separation of the target events from the other data.

In theory, PCA should allow for better separation of the clusters than simply plotting the raw data because the variance of the data set is preserved in a smaller number of dimensions. However, PCA clusters data together based on similarities that exist, but with no attempt to incorporate prior knowledge about the samples being analysed, and consequently PCA sometimes highlights effects that are not of direct relevance to the purpose of the experiment. This is described as an 'unsupervised' approach, and although this approach can be effective and may even be applauded in that it prevents the introduction of bias, it is very different to the approach that is normally taken in manual flow cytometric analysis. When manually analysing flow cytometry data the operator (or the scientist who has prepared the samples) will typically have prior knowledge of the cell type being analysed, preparation methods, etc., and good experimental design will necessitate that appropriate controls have been analysed to which experimental samples may be

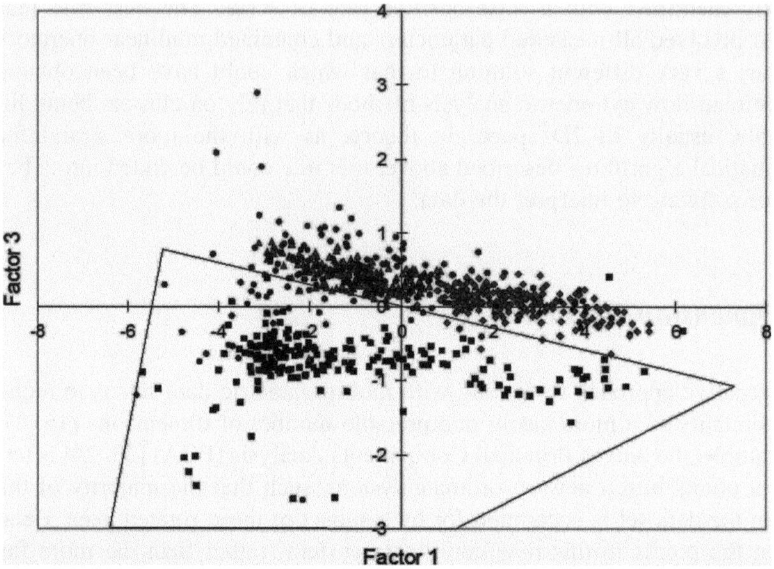

Fig. 9 PCA analysis of data set 1 showing discrimination of *B. globigii* spores (*squares*) from the other microorganisms. PCA was used to rotate the data points into a new co-ordinate system of discriminatory factors. A subset of the data representing all four organisms (200 examples of each organism) revealed that the first and third principal components gave the best partitioning of the data with 98.5% of *B. globigii* being correctly identified with 3.2% false positives

compared. This a priori knowledge can be incorporated into automated analysis methods via the use of a supervised learning method that relates input patterns from samples of known identity (control samples) to the desired output during the model development stage.

While unsupervised methods are ideal for a preliminary examination of the data [53], more sophisticated "supervised" methods such as Principal Components Regression (PCR), Partial Least Squares Regression (PLSR) [33] and Artificial Neural Networks (ANNs) [55] may produce better results. When using multiple variables as inputs to any multivariate analysis, some variables will be found to be more important than others. Indeed, it often happens that some variables are detrimental to the multivariate calibration model [25]. This could be because they are measuring something other than the searched-for correlation or simply because the information contained is also contained in other variables. The *Parsimony Principle* [45] states that where two models give the same result, the simpler model should be preferred, as it will be better at making predictions on an unseen data set.

4.3 Clustering Methods

As discussed earlier, manual data analysis seeks to identify clusters in the data, and a number of automated methods exist to do this. Clustering methods have recently been reviewed [57]. An algorithm based on multidimensional k-means clustering via a feature-guided approach has been reported [58] that accurately identified an artificial mixture of beads into the correct clusters. In the same year cluster analysis was shown to be better than PCA at discriminating between age classes of blood cell donors based on their T cell subsets [32]. This approach showed that bioinformatic methods could help to identify the cellular characteristics that best described and discriminated between a group of subjects—thus, this was not just an exercise in automatic clustering of data, but led to the development of new biological knowledge.

More recently, statistical model-based clustering approaches have been used to identify cell populations [31]. This work was based on a generalisation of Gaussian mixture models and thus was able to accommodate outliers. A direct multivariate finite mixture modeling approach has also been applied to flow cytometric data and has the advantage that it is suitable for detection of rare populations [42]. All of these methods seek to overcome the significant subjectivity and human time cost encountered in manual gating analysis, but none appear to have been adopted beyond the initial study.

4.4 Artificial Neural Networks

In contrast to PCA (see Sect. 4.2), Artificial Neural Networks (ANNs) represent a supervised learning method that exploits prior knowledge of class membership obtained from the analysis of samples of known identity [55]. ANNs were originally inspired by and modelled upon the biological neural networks in the central nervous system in an attempt to simulate through mimicry the logical intelligence of humans within a computer programme. The human brain is composed of interconnected neurons; each neuron is a specialised cell that can receive and transmit electrochemical signals. Neurons receive signals via a branching input structure of dendrites and pass signals on via their axons. Axons of one cell are connected to the dendrites of another cell via a synapse, and when activated via inputs to the dendrites the cell passes on an electrochemical signal via its axon. In this way the signal is transmitted via synapses to other neurons, which in turn may become activated. A neuron passes on a signal only if the total signal it receives from its dendrites exceeds a certain threshold; thus, the biological neural net is a highly flexible and responsive system. In an ANN, individual elements are similarly connected together to form a network. These elements, usually referred to as nodes of the network, have inputs and outputs, and through 'experiencing' different inputs together with known outputs the ANN identifies trends and is able to learn patterns and thus predict outputs from previously unseen combinations of inputs. In parallel with the biological neural network, the ANN is thus composed of many highly interconnected processing elements (Fig. 10a) that work together to solve a particular problem. Both artificial and biological neural networks have enormous capacity to infer input-output relationships from extremely complex inputs (stimuli).

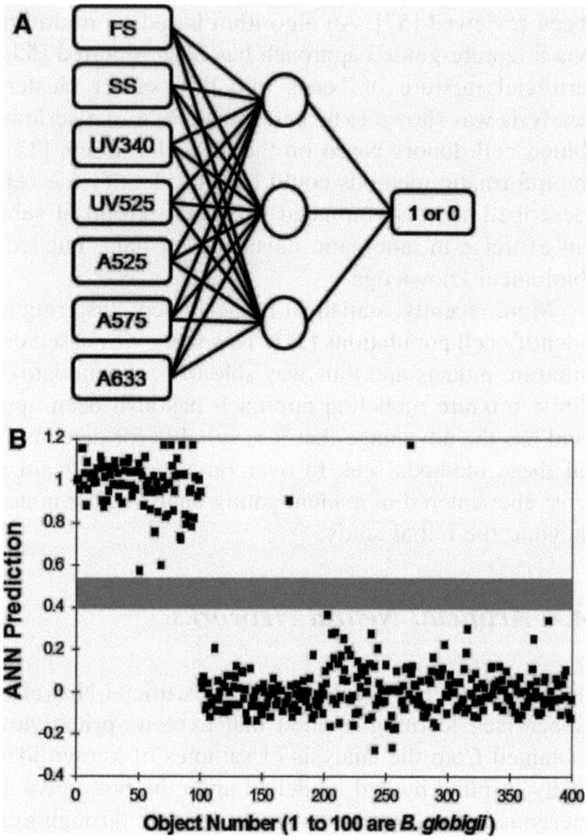

Fig. 10 Analysis of flow cytometric data using an ANN. **a** The ANN consists of an interconnected set of nodes. The input (*leftmost*) layer receives data representing each of the measured parameters (here fluorescence signals are tagged with 'UV' for signals originating from the HeCd laser and 'A' for those from the argon ion laser, together with the fluorescence wavelength). The ANN is trained, by repeated presentation of the training set, to associate *B. globigii* spores with an output value of 1 and all other cell types with a value of 0. Progress of training is assessed via the use of a separate data set, and weightings on the connections within the ANN are adjusted to reduce the error between the expected output and the predicted output. Each complete presentation of the training set is referred to as an epoch. **b** The trained ANN that was produced correctly identified all target organisms in a previously unseen data set with <1% false positives

In the case of flow cytometric data the problem may involve microbial species identification or the classification of events as corresponding to a particular type of cell or to a cell in a given physiological state. Trained ANNs have the ability to find meaning in complicated (e.g. multiparametric) or noisy data, and thus they can be used to infer a pattern or trend from a set of observations that would not be (immediately, if at all) apparent to the human eye and brain. Through the training process, the ANN becomes expert in the type of data that it has been exposed to.

Like human experts the ANN can then use its experience to make predictions from a novel set of inputs.

ANNs are also similar to the biological neural networks in that processing is carried out in parallel by the nodes rather than as sequential operations. There is no predefined algorithm (i.e. no preconceived ideas regarding which measured parameters are of most importance) and therefore no programmer input into which nodes become primarily responsible for a particular output, rather the network adapts to the training data. Nonetheless, in the majority of current software, although it is inspired by biology, a statistical and/or signal processing approach is frequently used for practical purposes—not least because there is so much about biological neural networks that we are still to learn.

Use of ANNs has been reported many times in flow cytometric data analysis (e.g. [4, 10, 56]), and many commercial packages and free programmes exist to analyse data sets. For ANNs to be applicable to a flow cytometric data set, there must be a relationship between the proposed known inputs (the parameters you measure on the flow cytometer) and the answer you are trying to obtain. It is acceptable that this relationship may be noisy, and often this will be the case due to both the inherent variability of biological material and the electronic and background noise present in the flow cytometer. Typically, however, at the outset the exact nature of the relationship between inputs and outputs is unknown—if it was known, it could be modelled directly rather than needing to be obtained through the supervised training of an ANN.

Before an ANN can be used to identify unknown organisms from cytometric data, it first needs to be trained to identify the organisms of interest. The exact method for data presentation will depend on the package chosen, but typically the data must first be converted from FCS format to ASCII as described for GPs above. To train an ANN with flow cytometric data, training and test data are prepared in a similar way to the process used for GP analysis in Sect. 4.1. For identification of a specific microorganism (e.g. *B. globigii* in data set 1) the inputs are a pattern of channel numbers corresponding to each of the measured parameters for the target organism. Additional patterns of inputs that represent other organisms (in this case *E. coli*, *M. luteus* and *S. cerevisiae*) must also be provided as negative controls. However, rather than the labels used in the GP the desired outputs are coded as 1 (for the target organism) and 0 for all other organisms in the training/test data sets. Furthermore, ANNs typically have the requirement that all inputs must be scaled between 0 and 1, but because of the sigmoidal activation function used in an ANN "headroom" must be available, and thus in the example presented here the data were scaled between 0.2 and 0.8. The neural network was taught, by repeated presentation of the training set, to associate inputs measured for *B. globigii* with an output of 1, and inputs for all other organisms were associated with an output of 0. In order to obtain the correct output, the 'weights' of each unit in the ANN are adjusted automatically to reduce the error between the desired output (one or zero) and the actual output (the output that the ANN at its current level of training is actually producing). The back propagation algorithm is the most widely used method for carrying out this process, which is performed without user intervention. Once the ANN is trained, it can be supplied with inputs of the channel numbers corresponding to unknown organisms, and its output (proximity to 0 or 1) will indicate if it is the organism of interest or not. In this way a trained ANN was produced after 100,000 epochs (presentations of the training data) and on interrogation correctly identified 100% of *B. globigii* with less than 1% false positives (Fig. 10b).

5 Cautionary Tales

Although a number of useful, automated methods for the analysis of flow cytometric data have been described here, it is worth stressing the need for care in applying them for routine analysis. For example, the consequences of "Shapiro's Seventh Law of Flow Cytometry" [46], which states that "No data analysis technique can make good data out of bad data", may be easily observed through the manual analysis of dot plots, but an automated system that leads to a number or a decision (e.g. live vs. dead) may not immediately ring equivalent alarm bells. Thus, quality control and the contemporary analysis of appropriate calibration and control samples are of paramount importance. Although many of the methods we have described are designed to be tolerant of noisy data, it is still best to avoid this wherever possible and certainly application of these methods does not provide an excuse to use a poorly maintained flow cytometer.

In generating training and test data, it is important that these are representative of the unclassified data that are to be analysed by the trained ANN/GP, etc. Thus, it is essential to ensure that no settings are changed on the flow cytometer and that the instrument has not drifted between generation of training and experimental data. Often, day-to-day variation in instrument performance may mean that it is necessary to produce a new model for the analysis of that day's data. Whilst this is becoming less problematic now that computer speed has increased, it is still time consuming in terms of manipulating data files.

In producing a trained ANN, for example, it is important that the net learns to predict unknowns rather than to learn the identity of each example in the training set. In the latter case the ANN loses its ability to generalise. Likewise, it is important that the data are collected in such a way that the differences between the controls used to generate the training data are due only to the expected difference between the samples. An often-quoted example of where a trained ANN failed to give the correct answer when presented with previously unseen data involved an attempt by the military to detect tanks in a forest. However, unwittingly the photographs with tanks present had been taken in different weather conditions to those of the "background" forest, and thus the ANN had learnt to identify the weather conditions rather than the desired information [41]. An equivalent flow cytometric problem could arise if a mistake was made in addition of a stain, for example, and whilst this would probably become apparent during manual analysis, it may not be apparent until a trained ANN failed to give the expected outputs.

Thus, the best advice that can be given is to use these tools where appropriate, but also to check the data in the time-honoured manual fashion.

6 Conclusions

The data analysis techniques discussed in this chapter offer a variety of methods for quickly and easily interpreting multiparametric flow cytometry data—reducing the data to a simple 1 or 0 ("yes" or "no"), rather than requiring the inspection of multiple dual parameter plots. Notwithstanding this, the majority of flow cytometric publications and applications rely upon manual data analysis by skilled scientists. In part this may be due to lack of familiarity with techniques and the different software packages required to undertake off-line analysis. In the future these techniques may have more widespread acceptance in the flow cytometry community if these advanced data analysis methods can be embedded in commercial or other dedicated flow cytometry software. Indeed, it has been noted that the multiparametric data sets that can be rapidly accumulated in flow cytometric experiments require efficient data viewing and complex clustering that could only be achieved using software specifically dedicated to flow cytometry [30]. Furthermore, in contrast to the recent advances in the technology that underpins flow cytometry, there has not been concomitant progress in the data analysis tools required to analyse, display and disseminate the data, and manual analysis of individual samples has been described as a limiting aspect of the technology [31]. It would appear that the time is right for a step forward to be made in flow cytometry data analysis methods, and it is hoped that the above chapter, together with the developmental FlowDec software, will facilitate this.

References

1. Avery SV (2006) Microbial cell individuality and the underlying sources of heterogeneity. Nat Rev Microbiol 4:577–587
2. Baumgarth N, Roederer M (2000) A practical approach to multicolor flow cytometry for immunophenotyping. J Immunol Methods 243:77–97
3. Bigos M, Baumgarth N, Jager GC et al (1999) Nine color eleven parameter immunophenotyping using three laser flow cytometry. Cytometry 36:36–45
4. Boddy L, Wilkins MF, Morris CW (2001) Pattern recognition in flow cytometry. Cytometry 44:195–209
5. Bravata DM, Zaric GS, Holty JEC et al (2006) Reducing mortality from anthrax bioterrorism: strategies for stockpiling and dispensing medical and pharmaceutical supplies. Biosecur Bioterror-Biodef Strategy Pract Sci 4:244–262
6. Chattopadhyay PK, Hogerkorp CM, Roederer M (2008) A chromatic explosion: the development and future of multiparameter flow cytometry. Immunology 125:441–449
7. Clark AJ, Petty HR (2008) Observation of calcium microdomains at the uropod of living morphologically polarized human neutrophils using flash lamp-based fluorescence microscopy. Cytometry Part A 73A:673–678

8. Dando M (1994) Biological warfare in the 21st century. Brassey's, London
9. Davey HM (2010) Prospects for the automation of analysis and interpretation of flow cytometric data. Cytometry Part A 77A:3–5
10. Davey HM, Jones A, Shaw AD et al (1999) Variable selection and multivariate methods for the identification of microorganisms by flow cytometry. Cytometry 35:162–168
11. Davey HM, Kell DB (1996) Flow cytometry and cell sorting of heterogeneous microbial populations—the importance of single-cell analyses. Microbiol Rev 60:641–696
12. Davey HM, Kell DB (1997) Fluorescent brighteners: novel stains for the flow cytometric analysis of microorganisms. Cytometry 28:311–315
13. Davey HM, Kell DB, Weichart DH et al (2004) Estimation of microbial viability using flow cytometry. In: Current Protocols in Cytometry (Supplement 29). Wiley, New York, pp 11.13.11–11.13.21
14. Day JP, Kell DB, Griffith GW (2002) Differentiation of *Phytophthora infestans* sporangia from other airborne biological particles by flow cytometry. Appl Environ Microbiol 68:37–45
15. Demers S, Kim J, Legendre P et al (1992) Analyzing multivariate flow cytometric data in aquatic sciences. Cytometry 13:291–298
16. Donalies UEB, Nguyen HTT, Stahl U et al (2008) Improvement of *Saccharomyces* yeast strains used in brewing, wine making and baking. Adv Biochem Eng Biotechnol 111:67–98
17. Frischmann U, Müller W (2006) Nine fluorescence parameter analysis on a four-color fluorescence activated flow cytometer. Cytometry Part A 69A:124–126
18. Fry B (2008) Visualizing data: exploring and explaining data with the Processing environment. O'Reilly Books, Sebastopol
19. Haefner JW (2005) Modeling biological systems: principles and applications. Springer, New York
20. Hammes F, Egli T (2010) Cytometric methods for measuring bacteria in water: advantages, pitfalls and applications. Anal Bioanal Chem 397:1083–1095
21. Hewitt CJ, Nebe-Von-Caron G (2004) The application of multi-parameter flow cytometry to monitor individual microbial cell physiological state. Adv Biochem Eng Biotechnol 89:197–223
22. Holty JEC, Kim RY, Bravata DM (2006) Anthrax: a systematic review of atypical presentations. Ann Emerg Med 48:200–211
23. Hotelling H (1933) Analysis of a complex of statistical variables into principal components. J Educ Psychol 24:417–441 and 498–520
24. Jolliffe IT (1986) Principal component analysis. Springer, Heidelberg
25. Kell DB, Sonnleitner B (1995) GMP-good modelling practice. Trends Biotechnol 13:481–492
26. Klanten R, Bourquin N, Ehmann S et al (2008) Data flow: visualising information in graphic design. Die Gestalten Verlag
27. Kosugi Y, Ikebe J, Shitara N et al (1986) Graphical presentation of multidimensional flow histogram using hexagonal segmentation. Cytometry 7:294–294
28. Lee JA, Spidlen J, Boyce K et al (2008) MIFlowCyt: the minimum information about a flow cytometry experiment. Cytometry Part A 73A:926–930
29. Lim DV, Simpson JM, Kearns EA et al (2005) Current and developing technologies for monitoring agents of bioterrorism and biowarfare. Clin Microbiol Rev 18:583–607
30. Lizard G (2007) Flow Cytometry analyses and bioinformatics: interest in new softwares to optimize novel technologies and to favor the emergence of innovative concepts in cell research. Cytometry Part A 71A:646–647
31. Lo K, Brinkman RR, Gottardo R (2008) Automated gating of flow cytometry data via robust model-based clustering. Cytometry Part A 73A:321–332
32. Lugli E, Pinti M, Nasi M et al (2007) Subject classification obtained by cluster analysis and principal component analysis applied to flow cytometric data. Cytometry Part A 71A:334–344
33. Martens H, Næs T (1989) Multivariate calibration. Wiley, Chichester

34. McLaughlin BE, Baumgarth N, Bigos M et al (2008) Nine-color flow cytometry for accurate measurement of T cell subsets and cytokine responses. Part I: Panel design by an empiric approach. Cytometry Part A 73A:400–410
35. Müller S, Davey H (2009) Recent advances in the analysis of individual microbial cells. Cytometry 75A:83–85
36. Müller S, Nebe-von-Caron G (2010) Functional single-cell analyses: flow cytometry and cell sorting of microbial populations and communities. FEMS Microbiol Rev 34:554–587
37. Murphy RF, Chused TM (1984) A proposal for a flow cytometric data file standard. Cytometry 5:553–555
38. Nebe-Von-Caron G (2009) Standardization in microbial cytometry. Cytometry Part A 75A:86–89
39. Novo D, Perlmutter NG, Hunt RH et al (1999) Accurate flow cytometric membrane potential measurement in bacteria using diethyloxacarbocyanine and a ratiometric technique. Cytometry 35:55–63
40. Pedreira CE, Costa ES, Barrena S et al (2008) Generation of flow cytometry data files with a potentially infinite number of dimensions. Cytometry Part A 73A:834–846
41. Priddy KL, Keller PE (2005) Artificial Neural Networks: an introduction. SPIE Publications, Washington
42. Pyne S, Hu X, Wang K et al (2009) Automated high-dimensional flow cytometric data analysis. Proc Natl Acad Sci USA 106:8519–8524
43. Reas C, Fry B (2007) Processing a programming handbook for visual designers and artists. MIT Press, Cambridge
44. Roederer M, Moody MA (2008) Polychromatic plots: graphical display of multidimensional data. Cytometry Part A 73A:868–874
45. Seasholtz MB, Kowalski B (1993) The parsimony principle applied to multivariate calibration. Anal Chim Acta 277:165–177
46. Shapiro HM (2003) Practical flow cytometry, 4th edn. Wiley, New York
47. Shiffman D (2008) Learning processing. Morgan Kaufmann (Elsevier) Publishers, Burlington
48. Sinclair M (2009) Maps for these territories. Creative Review July (in press). http://www.creativereview.co.uk/backissues/creative-review/2009/july-2009/maps-for-these-territories
49. Spidlen J, Moore W, Parks D et al (2010) Data file standard for flow cytometry, version FCS 3.1. Cytometry Part A 77A:97–100
50. Stopa PJ (2000) The flow cytometry of *Bacillus anthracis* spores revisited. Cytometry 41:237–244
51. Sychra JJ, Bartels PH, Bibbo M et al (1978) Dimensionality reducing displays in cell image analysis. Acta Cytol 21:747–752
52. Terzidis K (2009) Algorithms for visual design using the processing language. Wiley, Indianapolis
53. Tukey JW (1977) Exploratory data analysis. Addison-Wesley, Reading
54. Want A, Thomas ORT, Kara B et al (2009) Studies related to antibody fragment (Fab) production in *Escherichia coli* W3110 fed-batch fermentation processes using multiparameter flow cytometry. Cytometry Part A 75A:148–154
55. Wasserman PD (1989) Neural computing: theory and practice. Van Nostrand Reinhold, New York
56. Wilkins MF, Boddy L, Morris CW et al (1999) Identification of phytoplankton from flow cytometry data by using radial basis function neural networks. Appl Environ Microbiol 65:4404–4410
57. Xu R, Wunsch D (2010) Clustering. Wiley, Chichester
58. Zeng QT, Pratt JP, Pak J et al (2007) Feature-guided clustering of multi-dimensional flow cytometry datasets. J Biomed Inform 40:325–331

Adv Biochem Engin/Biotechnol (2011) 124: 211–227
DOI: 10.1007/10_2010_79

Published Online: 12 November 2010

From Single Cells to Microbial Population Dynamics: Modelling in Biotechnology Based on Measurements of Individual Cells

Thomas Bley

Abstract The development of dynamic modelling of microbial populations in bioprocesses is reviewed. In the 1960s Arnold Fredrickson established the theoretical basis of such models, and other researchers have subsequently advanced them. This review explores the relationships that describe cell proliferation and evaluates the importance of the application of flow cytometry to the fundamental parameterisation of the models for their use in bioprocess engineering. The section "Individual-Based Modelling" discusses recent theoretical developments. Delay-differential equations are demonstrated to describe accurately temporal variation of the cell proliferation cycle and specialised approaches and related iconography are applied to stochastic and deterministic modelling of stages of cellular development. Synchronised cultures of the bacterium *Cupriavidus necator* were prepared and monitored using a flow cytometer. The data obtained demonstrate that cell proliferation could be simulated quantitatively using the developed model.

Keywords Microbial population dynamics · Modelling · Flow cytometry · Synchronisation

Contents

T. Bley (✉)
Institute of Food Technology and Bioprocess Engineering,
Dresden University of Technology, 01062 Dresden, Germany
e-mail: thomas.bley@mailbox.tu-dresden.de

1 Introduction

To model the behaviour of microbial populations at the cellular level, criteria are required with which to distinguish one cell from another. The morphological and physiological differences between cells have a molecular basis, since the amounts and locations of all intrinsic molecules define the status of each cell. These parameters span a multi-dimensional state space, and the development of a cell can be described as a trajectory through this state space.

In order to design models for practical use, especially in biotechnology for bioprocess control and optimisation, the number of dimensions of the state space must be radically reduced by a mathematical projection of the original space onto sub-spaces. The modern biological equivalent of such a projection is the investigation of cells at the genomic, transcriptomic, proteomic, or metabolomic level.

The transition of cells or better the instant switch between different states is caused by either an internal (maturation) program or a change in the environment of the cell. The primary focus of this paper is the basic process of ageing, i.e., the progress of a single cell through the cell division cycle. This is indeed a fundamental process, as proliferation (and variation of the offspring) of individuals seems to be the most important feature for the evolution of life forms.

The mathematical tools that describe these processes map all the stages of the proliferation process into a compatible state space, which may be continuous or discrete and will be uni-dimensional if they map only the age of the cell, but multi-dimensional if more features, such as cell mass or cell functions, are described.

This study includes the review and discussion of historical papers from the field of dynamic population modelling and demonstrates that the mathematical tools for individual-based modelling are well established, but require application to the high-resolution analysis of single microbial cells to realise their true potential for bioprocess analysis and control.

2 The Classical Models

The first valid model of population heterogeneity was constructed by von Foerster [56] in a state space defined by the ages of the individuals. Cell age was interpreted as an index of the degree of maturation of a cell and, in later iterations, the maturation process of microbial cells was linked closely to their progress through the cell cycle.

Trucco [55] subsequently published a comprehensive and detailed discussion of this approach, and it was briefly, lucidly reviewed in German by Kiefer [31].

During the 1960s, the new scientific discipline of biochemical (or bioprocess) engineering emerged. Scale-up from laboratory to industrial operation required accurate modelling of the processes and was initially addressed using similar techniques to chemical engineering: mass balances were calculated for systems that were assumed to be homogeneous, and the heterogeneity of microorganisms was neglected.

A group of biochemical engineers and mathematicians of the Chemical Engineering Department of the University of Minnesota, inspired by Arnold G. Fredrickson, subsequently opened new scientific horizons by realising that understanding of microbial life (in a bioreactor) must be based on a quantitative description of how individual cells change their physiological state in response to both an internal program and interaction with the environment. The publication "Statistics and Dynamics of Procaryotic Cell Populations" [21] is widely recognised to be a key paper and to form the basis of all individual-based models in microbial population dynamics. The concept of a physiological state vector was clear, powerful, and could be represented by a matrix of concentration vectors to describe the internal structure of cells.

This rigorous concept allowed the classification of models using two pairs of antonyms, structured/unstructured and segregated/unsegregated, which are directly applicable to systems analysis in bioengineering [19]. The matrix representation of this analysis described by Bailey and Ollis [2] should be recommended reading for bioengineers.

As the terms segregated/unsegregated are not very evocative, the alternative terms corpuscular/non-corpuscular—or cellular/homogeneous [9]—were proposed later and were preferred by Fredrickson [53].

This generic model was applied to describe specific features of growth and proliferation at a single-cell level. Prime examples of this application were reported in two papers from the Bailey research group, in which a model was formulated using the Cooper-Helmstetter and Donachie hypotheses. These hypotheses attempt to explain the organisation of DNA reduplication in fast-growing prokaryotes, where the state space is defined by the cell mass and the cellular DNA content [45]. Other published analyses, also incorporating frequency functions for cell mass and DNA content, addressed eukaryotic cells in the form of the budding yeast *Saccharomyces cerevisiae* and included the interesting phenomena of asymmetric division and the accumulation of bud scars in the mother cells [29].

Modelling of the cell cycle of the baker's yeast *S. cerevisiae* stimulated the development of quantitative microbial population dynamics modelling, which was strongly supported by developments in flow cytometry, especially for the examination of continuous cultivation processes. The first description of population dynamics of *S. cerevisiae* using an age-structured model in parallel with flow cytometry was published in 1977 [51].

The phenomenon of autonomous oscillations in continuously cultivated *S. cerevisiae* is related to the partial synchronisation of cell division in the population. The discovery of this phenomenon created impetus for the development of the segregated models required to describe this process quantitatively.

The Alberghina research group based in Milan, Italy, published one of the first studies that combined the flow cytometric investigation of yeast cell populations with a quantitative description of dynamic cell processes using structured segregated models [13]. A collaboration between members of this group and Christos Hatzis later led to the development of morphologically structured models to describe the complex, heterogeneous structure of asymmetrically dividing cell populations [14, 27]. A recent paper by Porro [47] provides a comprehensive introduction to the theoretical understanding of cell size and structure in yeast populations.

Hjortso and colleagues have attempted to describe the phenomenon of oscillatory modes in continuous processes using an age distribution population balance model [30] and by numerical simulation [62], with which multiple oscillatory attractors could be reproduced. However, in the model simulations the attractors occurred at different dilution rates, while the oscillatory modes were observed experimentally at the same dilution rate.

Zhu et al. [63] combined age distribution modelling with mass balances to identify variables that can be compared directly to easily measured extracellular parameters. Mhaskar, Hjortso, and Henson [39] adopted this approach for the estimation of specific parameters needed for numerical simulations.

Bellgardt [3] developed a simple, age-structured model that included the description of asymmetrical division of baker's yeast cells, which quantitatively described the fluctuation of different cell types in sustained synchronous oscillation in yeast populations in continuous processes and advanced the modelling of population dynamics to practical bioprocess engineering. Bellgardt considered the quality of the final yeast product (e.g., the ability to produce carbon dioxide) to be influenced by the stage of the cellular propagation cycle. The fraction of budding cells was used successfully as a product quality index for feedback control of the process.

The Reuss research group explored the combination of a structured model in single cells (restricted to the sugar uptake system) with a segregated modelling approach to describe the behaviour of microbial populations in heterogeneous bioreactor systems and to calculate and visualise the heterogeneous distribution of metabolites in single cells caused by local glucose inputs and changing oxygen concentrations in a bioreactor [34, 35].

A fundamental limitation of individual-based modelling of bioprocesses is that it is usually not possible to measure cell state distributions in the system. In an illuminating paper entitled "Cytometric data as the basis for rigorous models of cell population dynamics", Srienc pointed out that "A complete understanding of the growth processes of a culture must be based on a quantitative understanding of how individual cells change in time and of how individual cells interact with the environment" [53]. Srienc discussed the classical Fredrickson [21] approach concisely and elaborated the primary requirement for practical use of the model: the accurate estimation of cellular parameters. Because the model requires information about a continuous state space, Srienc concluded that (flow) cytometric measurements are inevitably the most suitable source of such data. Modern molecular biological tools extend the determination of the physiological state of a cell far beyond cell size or DNA content to allow the detailed evaluation of cell composition, but there is a fundamental problem. A cell does not move continuously through its state space. Life is characterised by switching between the different, relatively stable stages that can be observed using cytometric investigation, and the cell cycle is the most important sequence of states in the life of a cell. Consequently, Fredrickson and Mantzaris [22] concluded that a population balance equation and a transition intensity function are required for each phase.

3 Some New Trends

The phenotypic cell-to-cell variation observed in isogenic microbial populations under the same environmental conditions presents a challenge to the dynamic modelling of microbial populations. It is necessary to elucidate the reason for this phenomenon and to find a mathematical tool to map it into an appropriate calculus. Mantzaris has postulated that there are two fundamentally different sources for this variability: unequal partitioning of cellular material during cell division and stochastic fluctuations associated with intracellular reactions. The combination of continuous and discrete variables has generated a deterministic cell population balance formulation and a fully stochastic Monte Carlo model, which can account for both intrinsic and extrinsic population heterogeneity sources. On the basis of simulation experiments, a very important finding was that simulation at the level of a single cell cannot lead to a comprehensive understanding of the stochastic behaviour of a population. This is because there is a complex interaction between single-cell genetic architecture and behaviour at the cell population level [38].

The generation of complex nonlinearities from the emergence of age classes is discussed further by Lavric and Graham [36].

Using a simplified model, Lee et al. [37] examined the influence of growth and substrate utilisation kinetics on population behaviour.

The Ramkrishna group conducted a detailed investigation of phase transition kinetics and parameterisation of age-structured models using experimental data from the evaluation of human leukaemia cells. The methodology thus developed is generically applicable to the determination of age-specific transition rates between cell cycle phases during balanced growth [50].

A recent and significant development is the consideration of the spatial structure, or living sphere of the microbial population, which is especially necessary to understand the behaviour of biofilms and the development of microbial colonies.

This aspect is not discussed in depth in this review, but some important elements for bioprocess engineering are highlighted: For example, iDynomics, the successor of BacSim, is a flexible, XML-based simulator for individual-based modelling of bacterial colony and biofilm growth [32], which has been demonstrated to model biofilms in a realistic manner [33], but the parameterisation of such models is complicated because mass balance is very difficult to monitor in such heterogeneous systems.

A further development introduced by physicists to describe the formation of co-operative growth patterns in bacterial colonies is a model that incorporates individual walkers (representing bacteria) that move in response to gradients in nutrient concentration and communicate with each other by means of a chemotactic "feedback" (quorum sensing). The self-organisation of such colonies can be described accurately using a simplified approach [4].

Also elementary altruistic behaviour was demonstrated by Xavier and Foster [61] in an individual-based model simulating the outcome of evolutionary competition between cells with a different level of extracellular polymer production.

The growth of biotechnologically relevant dimorphic yeast colonies under famine nutrient supply has been investigated [12], and a hybrid cellular automaton model has been developed. Further experimental parameterisation of this model has been achieved [57–59].

Cellular automaton models have been also developed to analyse spatio-temporal pattern formation in interacting cell populations, e.g., in populations of the ascomycete *Neurospora crassa* [17, 18].

The cellular Potts Model (CMP), which can also be regarded as an individual-based approach, has been used to discover a new mechanism that can explain the formation of aligned cell clusters in myxobacteria. This mechanism does not depend on cell cooperation, and in particular it does not depend on diffusive signals guiding cell motion [54].

In ecological modelling sciences the term "Individual-based Modelling" has been used for around 20 years. In a monograph the philosophy and the methods of this approach are described and discussed [24].

Based on this approach several studies have been conducted to demonstrate that the principles could also be transferred to the behaviour of microorganisms [20, 28, 48]. Surprisingly, all the fundamental scientific work done by bioprocess engineers in this field was not considered and reflected by these authors. Many passages of their publications remind one of "reinventing the wheel".

4 Stage Structure Models: The Geometry of Microbiological Time

It is evident that models of population dynamics for use in bioprocess engineering should be as simple as practicable and that their structure should be determined by measurements of the population distribution between the different states. Flow cytometry is the technique best suited to such measurement.

Investigations of budding yeast using cell cycle stage determination reveals two morphologically distinct stages: single cells and budding cells. They are connected via the four eukaryotic cell cycle stages: G1, S, G2, and M, which are determinable by measuring the cellular DNA content.

It has been demonstrated that simple, two-stage models can describe the phenomenon of auto-synchronisation in continuous bioprocesses. A requirement for describing such dynamic behaviour is model equations in which the stages have differing yields and maintenance coefficients, and the transition rates between the cell stages are influenced by the concentration of a limiting substrate. Such a model, using ordinary differential equations (ODEs), was published by Bley and Schmidt [5], and analytical investigation of the equations demonstrated that the emergence of a periodic attractor was caused by a Hopf bifurcation [8].

A similar model was successfully applied to the simulation of optimised PHB synthesis by *Methylobacterium rhodesianum* [1] by mapping two stages of the bacterial cell cycle into two variables that described different kinetics of product synthesis. The population dynamics were determined by flow cytometric analysis of the DNA and PHB content of the bacterial cells, and the data were applied to parameter identification [25].

However, ODEs cannot describe the structure of time for individual biological entities, because it is characterised by states of variable duration that depend on environmental conditions and phases of a determined duration [60]. The duration of the phases is almost constant and only slightly influenced by the environment, for example, the life cycle of an insect, the duration of human pregnancy, and the cell cycle of microorganisms.

This very basic property of cell cycle regulation was formulated as a hypothesis by Smith and Martin [52]. Thus, the kinetics of transition to the next state can be assumed to be stochastic, or there may be a time-controlled stage in which specific events occur stepwise, i.e., which is deterministic.

In the case of eukaryotic cells, the stochastic state corresponds to the G1 phase of the cell cycle, while the deterministic state corresponds to the S and M phases. By expansion of the Smith and Martin hypothesis, the G2 phase can be a second stochastic state.

In age-structured models, the existence of such a deterministic phase is mapped into the shape of the function that describes the dependency upon cell age of the division probability. Whereas in most applications this function is a Gaussian normal distribution, the Smith and Martin hypothesis provides an exponential distribution with a minimal age τ (Fig. 1).

Cooper developed a unifying model for eukaryotes and prokaryotes [15] in which the B (or I) phase represents the stochastic state and the C and D phases span the deterministic stage. Prokaryotes have developed a much broader spectrum of cell cycle regulation modes [43, 44], but the basic principle of a combination of deterministic and stochastic states seems to be universal.

In contrast to the regulation of cell proliferation in multi-cellular systems by substances produced internally by the organism, bioreactors depend upon control of the environmental conditions within the reactor in order to influence the behaviour of individual cells, for example, in response to the concentrations of substrates and nutrients. The transition rate from the stochastic to the deterministic state should be a function of nutrient or substrate concentration and can be described in a model by a kinetic rate function.

How should the behaviour of single cells or other individuals be mathematically described? The approach of Fredrickson and Mantzaris [22] is adequately generic

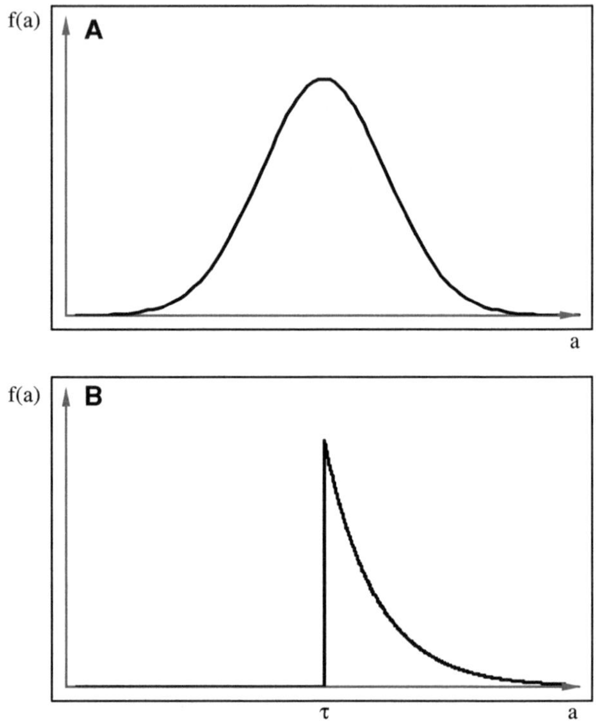

Fig. 1 Division probability function *f*(*a*). **a** Mathematically friendly Gauss function. **b** Reality with a minimum age τ for division

to map the principles of cell regulation, but it is highly theoretical and may not be practicable in general for parameter estimation. Thus, a mathematical model is required in which the variables describe the states that can be measured directly by a flow cytometer and, equally important, is clear to biologists.

Starting with an age distribution model, Gurney and Nisbet [26] derived a model using delay-differential equations (DDE) in which the delay term τ represents the duration of the deterministic stage. The model was demonstrated to describe real biological populations accurately—in this case, insect populations with larval and adult stages.

In parallel to the development noted above, the author of the present report and co-workers developed a related calculus, based upon the Smith and Martin hypothesis, to describe the microbial cell cycle and the population dynamics of the stages of the cell cycle in bioprocesses. This was significantly different to the Gurney and Nisbet approach because DDE calculus was applied to the deterministic stages, but ODE calculus was used for the stochastic states [6, 7].

Specialised icons were developed to depict the various patterns of regulation of proliferation. Stochastic states are mapped by a circle: 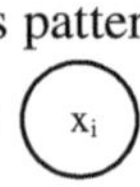

and deterministic stages by a rectangle: 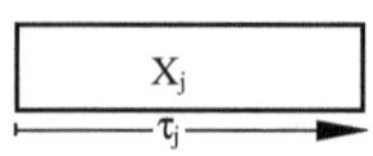

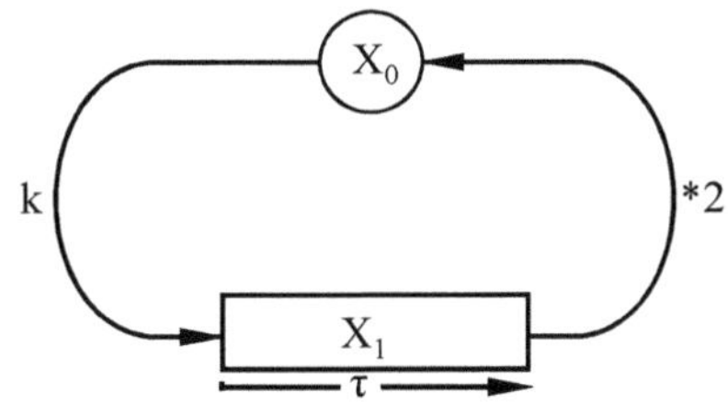

Fig. 2 Generic cell cycle graph

Figure 2 presents the generic structure of cell proliferation graphically. In this scheme the rate, *k*, of the transition from the stochastic to the deterministic state is constant, which means that it is not influenced by the environment. τ_j is the residence time in the deterministic stage.

The basic equation of this graph is a combination of ODE and DDE, and has the structure:

$$dx_0(t)/dt = 2\,kx_0(t-\tau) - k\,x_0(t)$$

$$dx_1(t)/dt = kx_0(t) - kx_0(t-\tau)$$

with the boundary conditions

$$x_0(t) = \psi(t), t[-\tau, 0]$$

$$x_1(t) = x_{1,\text{start}}$$

The boundary conditions determine the type of growth at the beginning of the simulation, which is either synchronised or balanced.

The specific growth rate, μ, for balanced growth depends on k and τ, and is calculated from the following transcendent equation:

$$\mu(k,\tau) = k\left(2\ e^{\mu(k,\tau)\tau} - 1\right).$$

Graphs of the regulation of proliferation can also be constructed for other microorganisms. The cell cycle of the budding yeast *Saccharomyces cerevisiae* is characterised by an asymmetric division, especially under poor cultivation conditions.

This is presented graphically in Fig. 3: x_0^{D} are the small daughter cells (G1 phase), x_0^{M} are the large daughter and the mother cells (G1 phase), and x_1 are proliferating cells (S/G2/M phase).

Another type of reproduction is demonstrated by the bacterium *Caulobacter crescentus* (Fig. 4), in which x_0^{SW} and x_1^{SW} are swarming cells, and x_1^{ST} are stalked, sessile cells.

The transition rates, k_i, in these graphs are constants. For coupling the cell proliferation process to the quality of the environmental conditions, especially the concentration of (limiting) nutrients or substrates, it is necessary to modify k_i to $k_i(s)$, which describes the probability of cells leaving the stochastic state as a

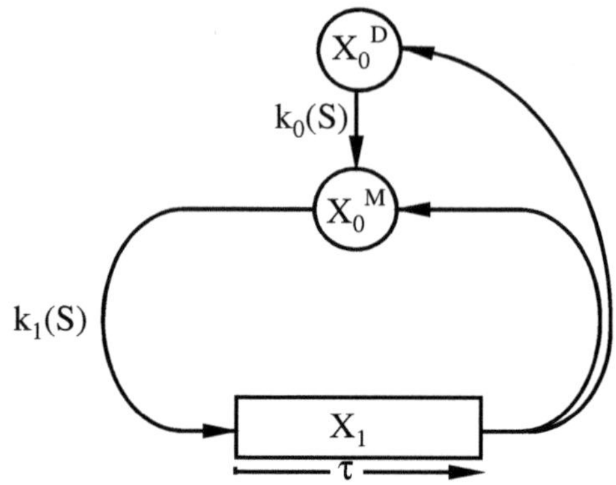

Fig. 3 Cell cycle graph of *Saccharomyces cerevisiae*. The transition rates $k_i(s)$ depend on the concentration of a limiting substrate s

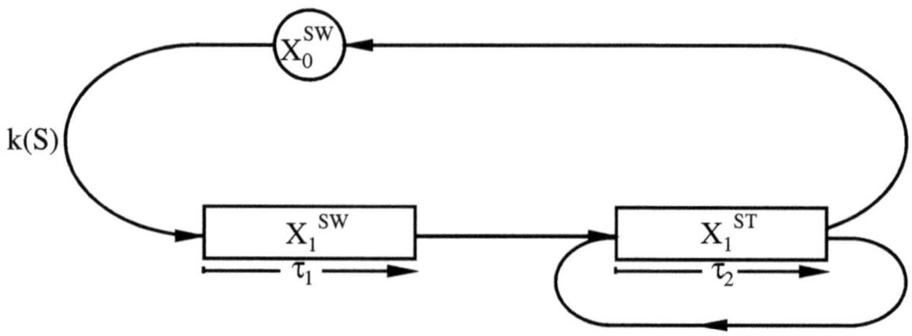

Fig. 4 Cell cycle graph of *Caulobacter crescentus*. There is a stalked (x^{st}) and a swarming (x^{sw}) form of these bacteria

function of substrate concentration. This function is typically a constantly increasing one, but it may also decrease.

Mueller described diverse types of cell cycle regulation in bacteria. One new finding and proposal is the so-called Pre-D phase in which cells pass over if the nutrition conditions are famine [43]. Typical bacteria with this proliferation pattern are *Acinetobacter calcoaceticus* and *Rhodococcus erythrophilus*. It is easy to construct the matching graph with the icon language; see Fig. 5. It is important to consider that $k_1(s)$ and $k_3(s)$ are monotonic increasing functions; in contradiction to this, $k_3(s)$ is a monotonic decreasing function.

The above modelling approaches were applied to describe phenomena such as the auto-synchronisation of yeast populations, and, using data from literature, the models could be parameterised [9, 10].

Theoretical investigations and practical experiments using these models verified that the application of periodic control regimes to continuous fermentation processes can produce higher specific yields, better performance, and improved specific process productivity [40, 41].

Unfortunately, the database was poor at this time because flow cytometry had not developed sufficiently to be applied to bioprocess engineering practically, and

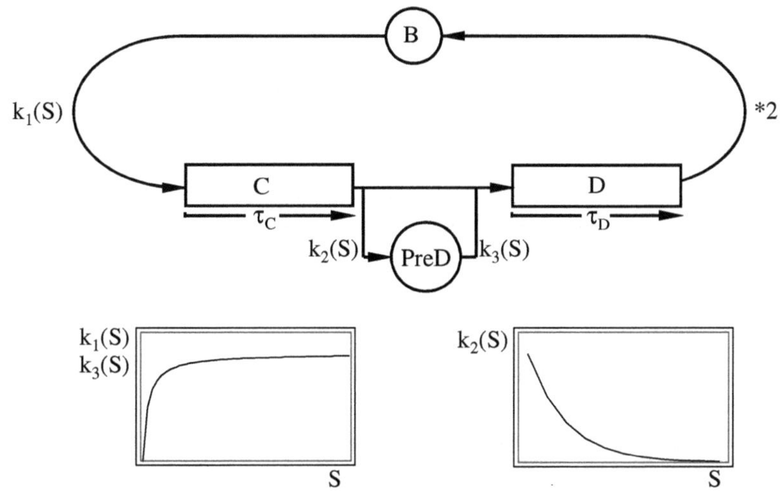

Fig. 5 Cell cycle graph of bacteria possessing a Pre-D phase, which is entered under famine culture conditions

the monitoring of population distributions in dynamic bioprocesses was not a major concern of microbiologists or bioprocess engineers.

5 Modelling Experiments with Synchronous Cultures

Fifteen years later, the research group of the present author and the Mueller research group in Leipzig re-evaluated the concept of phased cultivation that was developed by Dawson [16]. Phased cultivation is influenced by periodic changes between feast and famine growth conditions in the bioreactor, thus inducing synchrony [42]. By optimisation of the control parameters, it is possible to obtain synchronisation of the population, so that the whole population becomes strongly representative of the behaviour of a single cell, and single cell analysis, in a certain sense, thus can be replaced by analysis of the whole biomass in the bioreactor. Careful monitoring of the varying cell stage proportions is necessary to quantify the degree of synchronisation.

Fritsch et al. [23] reported the successful synchronisation of *Cupriavidus necator* under continuous phasing by managing the composition of the medium in such a way that the amount of the limiting substrate, pyruvic acid, was sufficient for a single duplication of the biomass (and the cell number). All other nutrients were balanced in surplus. Many methods have been published for the determination of the degree of synchronisation, including possibly the most objective method, developed by Priori and Ubezio [49], with which, using comprehensive flow cytometry analysis, a synchronisation percentage of 75.9% was estimated.

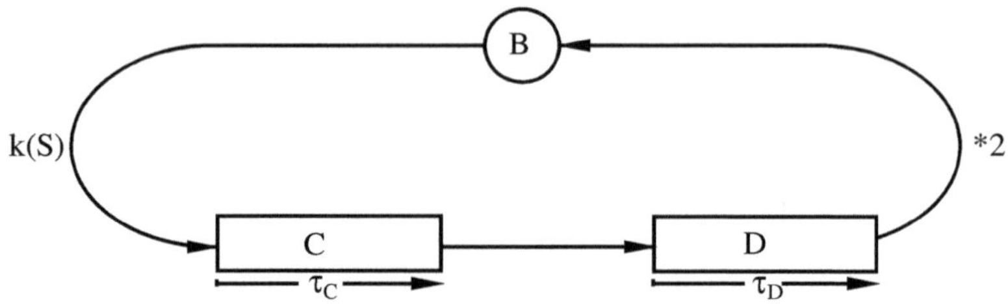

Fig. 6 Cell cycle graph of *Cupriavidus necator*

The degree of synchronisation was controlled by changing the duration of the phasing cycle.

According to Cooper [15], the cell cycle of *Cupriavidus necator* can be divided into three phases (Fig. 6). Cellular growth is associated with the B phase, whereas DNA synthesis and the following division into two daughter cells take place within the C and D phases of the cell cycle. The probability of the cell leaving this state depends on the quality of the extracellular environment (e.g., nutrient supply). In contrast, the C and D phases of the cell cycle form the deterministic state, which has a fixed traverse period, i.e., $\tau_{CD} = \tau_C + \tau_D$. Consequently, the growth rate of a population depends on the transition probability between these two states and the duration of the deterministic state. Hence, the doubling time t_d of the population equals the sum of τ_{CD} and the variable fraction of the B-phase duration.

On the basis of the already introduced delay-differential calculus, a process model for phased cultivation was developed [46].

$$\frac{\mathrm{d}B}{\mathrm{d}t} = 2k_{CD}(s)B(t-\tau_{CD}) - k(s)B(t) - \sum_{i=1}^{n_c} \alpha_N \delta(t-t_i)B(t)$$

$$\frac{\mathrm{d}C}{\mathrm{d}t} = k(s)B(t) - k_C(s)B(t-\tau_C) - \sum_{i=1}^{n_c} \alpha_N \delta(t-t_i)C(t)$$

$$\frac{\mathrm{d}D}{\mathrm{d}t} = k_C(s)B(t-\tau_C) - k_{CD}(s)B(t-\tau_{CD}) - \sum_{i=1}^{n_c} \alpha_N \delta(t-t_i)D(t)$$

$$\frac{\mathrm{d}s}{\mathrm{d}t} = -\frac{1}{Y_{N/s}}\mu(s)N(t) + \sum_{i=1}^{n_c} \alpha_s \delta(t-t_i)(2s_0 - s(t))$$

$$\delta(t-t_i) = \begin{cases} 0 & t \neq t_i \\ \infty & t = t_i \end{cases}$$

The draining and filling of a fixed reactor volume fraction α is modelled by a Dirac-function δ, i.e., generating a pulse after each phase cycle i.

A model simulation demonstrated that it was reasonable to describe the stop of all single cells in the B state after passing the C and D phase when the substrate was exhausted. Figure 7 presents a simulated example with a transition rate of

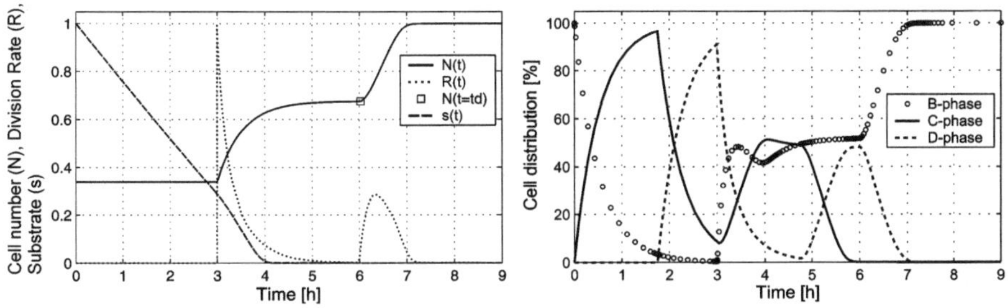

Fig. 7 Simulation of cells accumulating in the B phase under substrate limitation $B_0 = 10^6$ cells/l; $s_0 = 1$ g/l; $\tau_C = 1.75$ h; $\tau_{CD} = 3$ h; $k_{max} = 2$ h^{-1}; $\mu_{max} = 0.5$ h^{-1}; $K_s = K_T = 0.03$ g/l; $Y_{N/s} = 2 \times 10^6$ cells/g [46]

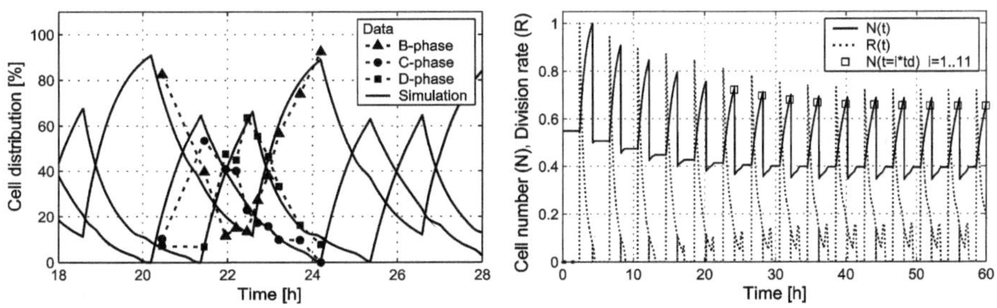

Fig. 8 Simulation of a phased culture of *Cupriavidus necator*. Phased cultivation was immediately started after a short batch phase at $t = 0.2$ h. $B_0 = 10^{11}$ cells/l; $s_0 = 0.85$ g/l; $\tau_C = 1.17$ h; $\tau_{CD} = 2.39$ h; $k_{max} = 1.07$ h^{-1}; $\mu_{max} = 0.73$ h^{-1}; $K_s = 0.003$ g/l; $K_T = 0.05$ g/l; $Y_{N/s} = 1.79 \times 10^{11}$ cells/g; $T = 4$ h [46]

$k = 2$ h^{-1} in which the population changes to asynchronous growth and the division rate converges to a mean value of $\bar{R} = \ln 2/t_d$.

This model could be successfully parameterised using data from flow cytometric investigations [23]. A genetic algorithm was applied to estimate the model parameters. The results of fitting the model to a flow cytometry dataset (6th phase cycle) with a defined phase interval of $T = 4$ h are presented graphically in the left panel of Fig. 8. The initial values for the B-phase cells (B_0) and the substrate concentrations of the exchange medium (s_0) were defined in accordance with the measurements. The model reproduced the flow cytometric measurements with high accuracy.

The simulation demonstrated that the culture reached a stable state, with a balance between nutrient supply and population size, resulting in exactly one doubling of the culture in each phase interval. Thus, with this really simple set of DDE, it is possible to describe the population dynamics in a tricky and complex controlled bioprocess. The reason for this success is due to a correct mapping of the geometry of microbiological time in the model.

6 Conclusion

Currently, individual-based models of microbial population dynamics are not applied to a significant extent in bioprocess optimisation and control. As it becomes increasingly obvious that the modelling of microbial performance can strongly influence the productivity and economic efficiency of bioprocesses, it is necessary to ask why there has been so little scientific progress in this field. It seems that incomplete communication between scientists who perform single cell analysis or modelling and bioprocess engineers is a potential major cause.

There are three good reasons to improve upon the under-utilisation of such models:

1. High-resolution microbial single cell analysis has advanced and can now provide quantitative analysis of the dynamic behaviour of microbial communities in bioreactors. These data provide an excellent basis for the parameterisation of mathematical models.
2. The fundamental model approaches to describe microbial population dynamics, originated mainly by the schools of Fredrickson and Bailey, can be translated into algorithms that are accessible to bioprocess engineers and bioscientists, and which facilitate computer simulation as well as the interpretation of simulation data.
3. Mass balance is the fundamental quantitative description of a bioprocess. It is known that the metabolite fluxes in different stages of microbial life are unequal, and recently it has become possible to cultivate single cells and observe their morphology and other features. However, it is currently impossible to establish mass balances for single cell cultivation.

The solution to this constraint is to synchronise the population in the bioreactor so that synchronous growth is used to couple single cell behaviour with mass balance. Thus, synchronisation is the link between single cell analysis and bioprocess engineering.

In 2002 the editors of this journal, Susann Mueller and Thomas Bley, asked the question “How should microbial life be quantified to optimise bioprocesses?” [11].

The answer is that all integral measurement methods that are available should be used. Without local measurement at the single cell level, it is not possible to obtain all the information necessary for efficient process control and optimisation, i.e., only the additional use of single cell analysis methods can provide a full picture. Cytomics will be the key that opens the door to more comprehensive understanding of bioprocesses.

Acknowledgments The author would like to thank Florian Centler, Andreas Deutsch, and Felix Lenk for their critical reading of the manuscript and their helpful notes.

References

1. Ackermann J-U, Müller S, Lösche A, Bley T et al (1995) *Methylobacterium rhodesianum* cells tend to double the DNA content under growth limitations and accumulate PHB. J Biotechnol 39:9–20
2. Bailey JE, Ollis FD (1977) Biochemical engineering fundamentals. McGraw-Hill, New York
3. Bellgardt K-H (1994) Analysis of synchronous growth of baker's yeast. Part i: development of a theoretical model for sustained oscillations. J Biotechnol 35:19–33
4. Ben-Jacob E, Schochet O, Tenenbaum A, Cohen I et al (1994) Generic modeling of cooperative growth patterns in bacterial colonies. Nature 368:46–49
5. Bley T, Schmidt A (1980) A two state microbial growth model for continuous fermentation. Stud Biophys 78:11–12
6. Bley T, Heinritz B, Schmidt A (1984) Some stationary properties of a two-state microbial growth model for continuous fermentation derived from the smith and martin hypothesis. Stud Biophys 98:119–124
7. Bley T (1987) State-structure models of microbial growth. Acta Biotechnol 7:173–177
8. Bley T, Wegner B (1988) Hopf bifurcation for a family of two-state microbial growth models. Acta Biotechnol 8:267–275
9. Bley T (1990) State structure models—a base for efficient control of fermentation processes. Biotechnol Adv 8:233–259
10. Bley T (1992) Delay-differential equations for modeling synchrony and periodic phenomena in microbial population dynamics. In: Karim MN, Stephanopoulos G (eds) Modeling and control of biotechnical processes. Pergamon Press, NY, pp 195–199
11. Bley T, Müller S (2002) How should microbial life be quantified to optimize bioprocesses. Acta Biotechnol 22:401–409
12. Boschke E, Bley T (1998) Growth patterns of yeast colonies depending on nutrient supply. Acta Biotechnol 18:17–27
13. Cazzador L, Mariani L, Martegani E, Alberghina L (1990) Structured segregated models and analysis of self-oscillating yeast continuous cultures. Bioprocess Eng 5:175–180
14. Cipollina C, Vai M, Porro D, Hatzis C (2007) Towards understanding of the complex structure of growing yeast populations. J Biotechnol 128:393–402
15. Cooper S (1979) A unifying model for the g1 period in prokaryotes and eukaryotes. Nature 280:17–19
16. Dawson PSS (1972) Continuously synchronized growth. J Appl Chem Biotechnol 22:79–103
17. Deutsch A, Dress A, Rensing L (1993) Formation of morphological differentiation patterns in the ascomycete *Neurospora crassa*. Mech Dev 44:17–31
18. Deutsch A, Dormann S (2005) Cellular automation modelling of biological pattern formation. Birkhauser, Boston
19. Eakman JM, Fredrickson AG, Tsuchiya HM (1966) Statistics and dynamics of microbial cell populations. Chem Eng Prog 62:37–49
20. Ferrer J, Prats C, López D (2008) Individual-based modelling: an essential tool for microbiology. J Biol Phys 34:19–37
21. Fredrickson AG, Ramkrishna D, Tsuchiya HM (1967) Statistics and dynamics of procaryotic cell populations. Math Biosci 1:327–374
22. Fredrickson AG, Mantzaris NV (2002) A new set of population balance equations for microbial and cell cultures. Chem Eng Sci 57:2265–2278
23. Fritsch M, Starruss J, Loesche A, Mueller S et al (2005) Cell cycle synchronization of *Cupriavidus necator* by continuous phasing measured via flow cytometry. Biotechnol Bioeng 92:635–642
24. Grimm V, Railsback SF (2005) Individual-based modeling and ecology. Princeton University Press, Princeton

25. Große-Uhlmann R, Bley T (1999) A modular approach to situation identification of the dynamics of bacterial populations synthesizing poly-β-hydroxybutyrate. Bioprocess Eng 21:191–200
26. Gurney WS, Nisbet RM (1984) The systematic formulation of delay-differential models of age or size structured populations. Lect Notes Biomath 52:163–172
27. Hatzis C, Porro D (2006) Morphologically-structured models of growing budding yeast populations. J Biotechnol 124:420–438
28. Hellweger FL, Bucci V (2009) A bunch of tiny individuals—individual-based modeling for microbes. Ecol Model 220:8–22
29. Hjortso MA, Bailey JE (1982) Steady-state growth of budding yeast populations in well-mixed continuos-flow microbial reactors. Math Biosci 60:235–263
30. Hjortso MA, Nielsen J (1995) Population balance models of autonomous microbial oscillations. J Biotechnol 42:255–269
31. Kiefer J (1973) Zur Mathematischen Beschreibung der Zellproliferation. Biophysik 10:115–124
32. Kreft J-U, Booth G, Wimpenny JWT (1998) Bacsim, a simulator for individual-based modelling of bacterial colony growth. Microbiology 144:3275–3287
33. Kreft J-U, Picioreanu C, Wimpenny JWT, van Loosdrecht MCM (2001) Individual-based modeling of biofilms. Microbiology 147:2897–2912
34. Lapin A, Müller D, Reuss M (2004) Dynamic behavior of microbial populations in stirred bioreactors simulated with Euler–Lagrange methods: traveling along the lifelines of single cells. Ind Eng Chem Res 43:4647–4656
35. Lapin A, Schmidt J, Reuss M (2006) Modelling the dynamics of *E. coli* populations in the three-dimensional turbulent field of a stirred-tank bioreactor—a structured-segregated approach. Chem Eng Sci 61:4783–4797.
36. Lavric V, Graham DW (2010) Birth, growth and death as structuring operators in bacterial population dynamics. J Theor Biol 264:45–54
37. Lee MW, Vassiliadis VS, Park JM (2009) Individual-based and stochastic modelling of cell population dynamics considering substrate dependency. Biotechnol Bioeng 103:891–899
38. Mantzaris NV (2007) From single-cell genetic architecture to cell population dynamics: quantitatively decomposing the effects of different population heterogeneity sources for a genetic network with positive feedback architecture. Biophys J 92:4271–4288
39. Mhaskar P, Hjortso MA, Henson MA (2002) Cell population modelling and parameter estimation for continuous cultures of *Saccharomyces cerevisiae*. Biotechnol Prog 18:1010–1026
40. Möckel B, Bley T, Böhme B (1989) Cyclic control of continuous biotechnological processes on the basis of a hierarchical control system. Syst Anal Model Simul 6:181–196
41. Möckel B, Bley T, Böhme B (1990) Model simulation of an efficient periodic control strategy for continuous fermentation processes. Acta Biotechnol 10:395–400
42. Müller S, Bley T, Babel W (1999) Adaptive responses of *Ralstonia eutropha* to feast and famine conditions analyzed by flow cytometry. J Biotechnol 75:81–97
43. Müller S (2007) Modes of cytometric bacterial DNA pattern: a tool for pursuing growth. Cell Prolif 40:621–639
44. Müller S, Harms H, Bley T (2010) Origin and analysis of microbial population heterogeneity in bioprocesses. Curr Opin Biotechnol 21:100–113
45. Nishimura Y, Bailey JE (1980) On the dynamics of Cooper-Helmstetter-Donachie populations. Math Biosci 51:305–328
46. Noack S, Klöden W, Bley T (2008) Modelling synchronous growth of bacterial populations in phased cultivation. Bioprocess Biosyst Eng 31:435–443
47. Porro D, Vai M, Vanoni M, Alberghina L et al (2009) Analysis and modeling of growing budding yeast populations at the single cell level. Cytometry 75A:114–120
48. Prats C, López D, Giró A, Ferrer J et al (2006) Individual-based modelling of bacterial cultures to study the microscopic causes of the lag phase. J Theor Biol 241:939–953

49. Priori L, Ubezio P (1996) Mathematical modelling and computer simulation of cell synchrony. Methods Cell Sci 18:83–91
50. Sherer E, Tocce E, Hannemann RE, Rundell AF et al (2008) Identification of age-structured models: cell cycle phase transition. Biotechnol Bioeng 99:960–974
51. Slater ML, Sharrow SO, Gart JJ (1977) Cell cycle of *Saccharomyces cerevisiae* in populations growing at different rates. Proc Natl Acad Sci USA 74:3850–3854
52. Smith JA, Martin L (1973) Do cells cycle? Proc Natl Acad Sci USA 70:1263–1267
53. Srienc F (1999) Cytometric data as the basis for rigorous models of cell population dynamics. J Biotechnol 71:233–238
54. Starruß J, Bley T, Sogaard-Andersen L, Deutsch A (2007) A new mechanism for collective migration in *Myxococcus xanthus*. J Stat Phys 128:269–286.
55. Trucco E (1965) Mathematical models for cellular systems: the von Foerster equation. Part i and ii. Bull Math Biophys 27:285–304 (see also pp 449–471).
56. von Foerster H (1959) Some remarks on changing populations. In: Stohlman F (ed) The kinetics of cellular proliferation. Grune & Stratton, New York
57. Walther T, Reinsch H, Große A, Ostermann K et al (2004) Mathematical modeling of regulatory mechanisms in yeast colony development. J Theor Biol 229:327–338
58. Walther T, Reinsch H, Ostermann K, Deutsch A et al (2005) Coordinated development of yeast colonies: B) quantitative modeling of diffusion-limited growth. Eng Life Sci 5:125–133.
59. Walther Th, Reinsch H, Ostermann K, Deutsch A, Bley Th (2010) Applying dimorphic yeasts as model organisms to study mycelial growth: Part 2: application of mathematical models to identify different construction principles in yeast colonies. Bioprocess Biosyst Eng. doi:10.1007/s00449-010-0443-5. Accessed 15 June 2010
60. Winfree AT (1990) The geometry of biological time. Springer, Berlin
61. Xavier JB, Foster KR (2007) Cooperation and conflict in microbial biofilms. Proc Natl Acad Sci USA 104:876–881
62. Zamamiri AM, Zhang Y, Henson MA, Hjortso MA (2002) Dynamics analysis of an age distribution model of oscillating yeast cultures. Chem Eng Sci 57:2169–2181
63. Zhu G-Y, Zamamiri A, Henson MA, Hjortso MA (2000) Model predictive control of continuous yeast bioreactors using cell population balance models. Chem Eng Sci 55:6155–6167

Index

MIX
Papier aus verantwortungsvollen Quellen
Paper from responsible sources
FSC® C105338

If you have any concerns about our products,
you can contact us on
ProductSafety@springernature.com

In case Publisher is established outside the EU,
the EU authorized representative is:
Springer Nature Customer Service Center GmbH
Europaplatz 3, 69115 Heidelberg, Germany

Printed by Libri Plureos GmbH
in Hamburg, Germany